William Tebb

The Recrudescence of Leprosy and Its Causation

A popular treatise

William Tebb

The Recrudescence of Leprosy and Its Causation
A popular treatise

ISBN/EAN: 9783337315405

Printed in Europe, USA, Canada, Australia, Japan

Cover: Foto ©berggeist007 / pixelio.de

More available books at **www.hansebooks.com**

THE
RECRUDESCENCE OF LEPROSY

AND ITS

CAUSATION

A Popular Treatise

By WILLIAM TEBB

WITH AN APPENDIX

LONDON
SWAN SONNENSCHEIN & CO
PATERNOSTER SQUARE
1893

LEPROSY is, perhaps, the most terrible disease that afflicts the human race. It is hideously disfiguring, destructive to the tissues and organs in an unusual degree, and is hopelessly incurable, the fate of its victims being, indeed, the most deplorable that the strongest imagination can conceive, and many years often passing before death rids the unhappy sufferer from a life of misery to which there is scarcely any alleviation. It is not to be wondered at, then, that the question is one which philanthropists in these enlightened days are taking up actively.—*British Medical Journal, Nov. 19, 1887.*

WHAT an appalling disease leprosy is; how it marks its victims and maims them; and how, through a prolonged period of suffering, it leads to a sure death.—*Trinidad Leprosy Report for 1889, page 75.*

THERE is no known remedy for the disease (leprosy). Prevention can alone cope with it.—*Lancet, April 27, 1889.*

THERE are few things which are more interesting, few which are better fitted to instruct while they humiliate, than an occasional retrospect of the fate which befalls new remedies or fresh measures which are ever and anon being introduced for the alleviation or cure of disease. Each has, as a rule, to pass through three distinct stages. The first is the stage of unreasoning enthusiasm, when much is said about a sovereign balm or a great advance in therapeutics, and when a pitying contempt is expressed for antiquated methods hitherto in use. After a little time a second stage is reached. The natural swing of the pendulum has come, and disillusion and disenchantment, with the irritation which these processes beget in the too credulous, take the place of unlimited praise and fulsome adulation. It is now discovered that the hitherto vaunted remedy is not only useless, but that it is positively harmful. *Lancet, September 12, 1891.*

THE fact that the leprosy may be inoculated, I consider to be proved as much as any fact in medical science. *Dr. R. Hall Bakewell, Physician to the Leper Asylum, Trinidad.*

THERE are only two foundations of law—utility and necessity—and they are both of them conditions without which nothing can give it any force—I mean equity and utility.—*Edmund Burke.*

DEDICATION.

TO THE

ROYAL COMMISSION ON VACCINATION,

WHOSE PROTRACTED AND PATIENT LABOURS HAVE ELICITED THE VALUABLE EVIDENCE CONTAINED IN ITS ALREADY PUBLISHED REPORTS,

THIS WORK

DEALING WITH AN IMPORTANT PHASE OF THEIR INQUIRY IS RESPECTFULLY DEDICATED BY THE

AUTHOR.

CONTENTS.

DEDICATION.

Preface, — — — — — — — — — PAGE 7

CHAPTER I.
The Increase of Leprosy, — — — — — — — 21

CHAPTER II.
Is Leprosy Contagious? — — — — — — — 80

CHAPTER III.
Leprosy Communicable by Inoculation, — — — — 97

CHAPTER IV.
Vaccination with Reference to Leprosy, — — — — 131

CHAPTER V.
Leprosy and Re-Vaccination, — — — — — 216

CHAPTER VI.
Other Alleged Causes of Leprosy, — — — — — 223

CHAPTER VII.
Inadequacy of Medical Theories of Causation, — — — 231

CONTENTS CONTINUED.

CHAPTER VIII.
Leprosy and Vaccination at the International Hygienic Congress, - - - - - - - - - 236

CHAPTER IX.
Vaccination Ignored in Official Leprosy Reports, - - 241

CHAPTER X.
Official Statistics, - - - - - - - - 256

CHAPTER XI.
Leprosy and the Aboriginal Races, - - - - - 264

CHAPTER XII.
Vaccinal Diseases in South Africa, - - - - - 268

CHAPTER XIII.
A Visit to the Lazaretto, Robben Island, South Africa, - 276

CHAPTER XIV.
The Segregation of Lepers, - - - - - - 282

CHAPTER XV.
Self-Devotion to Lepers, - - - - - - - 290

CHAPTER XVI.
The Leprosy Investigation Committee, - - - - 295

CHAPTER XVII.
Leprosy Incurable — Hygiene the only Palliative, - - 310

SUMMARY, · 350

CONTENTS CONTINUED.

	PAGE
APPENDIX,	353
Vaccination Frauds,	354
Compulsory Vaccination in Bombay,	355
The Revolt against Compulsory Vaccination in India,	360
Vaccination in the West Indies,	363
Vaccination a Failure and a Danger to Health from its Inception,	364
Vaccination Failures in 1817,	367
Arm-to-arm versus Calf-Lymph Vaccination,	367
Dr. R. H. Bakewell on the Risks of Vaccination,	371
Leprosy in Madeira,	373
Sir James Paget on Surgical Pathology,	375
Dr. M. D. Makuna's Vaccination Census,	376
Is Vaccination a Preventive of Small-Pox?	379
Medical Denials and Admissions,	380
How Leicester deals with Variolous Outbreaks,	382
Dr. Creighton on the Natural History of Cow-pox and Vaccinal Syphilis,	387
The Author's Personal Statement of the Results of Vaccination,	389
Index,	393
List of Works and Pamphlets referred to in foregoing pages,	401

CORRECTIONS.

Page 36.—For "*The Lancet*" read "Mr. Hamilton Cartwright, in *The Lancet.*"

,, 37.—Line 30, for "Bourmetz" read "Beaumetz."

,, 58.—Fifth line from bottom, for "in the third report of" read "by the Minutes of Evidence with the third report of."

,, 69.—For "Lock" read "Loch."

,, 99.—Third line from bottom, for "Pach" read "Pacha."

,, 129.—The "elephantiasis" conveyed by mosquitoes appears to have been another kind of disease than leprosy.

,, 131.—*Note.* Sir John Simon's words occur in his "English Sanitary Institutions," London, 1890, p. 123, *note*, as follows:—"I venture to predict that the new evidence [before the Royal Commission on Vaccination], so far as it may regard the merit of the discovery, will establish more firmly than ever that Jenner's services to mankind, in respect of the saving of life, have been such that no other man in the history of the world has ever been within measurable distance of him."

,, 187.—Line 4, for "address delivered," etc., read "paper read at the meeting of the British Medical Association."

,, 223.—Line 10, for "Armour" read "Armauer."

,, 224.—Line 4, for "Louis" read "Lewis."

,, 241.—Fourth line from bottom, and elsewhere, for "Bevan" read "Beaven."

Bacillus Lepræ for *Bacillus Lepra* when occurring.

PREFACE

THE remarkable spread of leprosy during the past thirty years has excited much public attention. Having for many years been interested in the public health, I have been prompted to investigate the causation of this increase. My attention was first called to the growing ravages of leprosy during a visit to Asia Minor in the year 1884, and to one source of infection the extent of which is as yet imperfectly realised (I mean vaccination), by the perusal of the evidence brought before the Select Parliamentary Committee on Vaccination of 1871 by Dr. R. Hall Bakewell. In 1888-89, during a visit to the Virgin, Leeward, and Windward Islands, British Guiana, and Venezuela, I took the opportunity of investigating this serious allegation. In the course of my inquiries I obtained particulars of a number of cases of leprosy due to vaccination. These were furnished by highly respectable colonists, but were often coupled with the request that no names were to be published, either of the suffering families or of those who communicated the details. This reluctance, which is entitled to every consideration, was due to the fear of exposing relatives, and damaging their social standing in the community where they reside. Although the danger of communicating leprosy by vaccination has been admitted in official and other reports, I have on many occasions found it extremely difficult to get at the facts, copies of important documents having been repeatedly refused by officials both at home and abroad, and notably in the French Colonial possessions.

It is perhaps too much to expect that those who regard Jenner as one of the greatest of human benefactors will display much energy in bringing to light such cases as I have referred to.

Some of the obstacles in the way of independent investigation of this subject are alluded to in a communication, read before the Royal Vaccination Commission, from Dr. Charles E. Taylor, Secretary to the Legislative Council, Island of St. Thomas, Danish West Indies. Dr. Taylor states that during his 20 years residence at St. Thomas, D.W.I., he has known many cases of the communication of leprosy by means of the vaccinator's lancet, but he adds that the sufferers or their families invariably decline to have the fact disclosed. One resident physician in Honolulu, who told me how the disease had been disseminated by means of vaccination in Hawaii, strongly deprecated making the details known, as he would not answer for the consequences. Despite these difficulties, however, a good many cases are recorded by medical men of high standing and wide experience, and some of them are here presented to the reader. Cases have come under my cognizance in which the reports of district surgeons, showing the spread of terrible diseases by means of vaccination, have been officially suppressed.

In June, 1890, I appeared as a witness before the Royal Commission on Vaccination, and gave evidence as to the results of my inquiries up to that date. This evidence will be found in the third report of the proceedings, pages 154-161.

At that time my investigations had been mainly limited to the West Indies, British Guiana, and Venezuela. Since then I have extended my personal investigations to Norway, California, the Sandwich Islands, Ceylon, Egypt, New Zealand, Cape Colony and Natal in South Africa, and most of the Colonies in Australia, and have put myself in communication with superintendents of leper asylums and leading dermatologists in all other countries where leprosy is endemic. The results of these inquiries, with other collateral evidence bearing on the subject, are briefly set forth in this volume. While the *pros* and *cons* of the theories of heredity, fish-eating, malaria, and contagion, have been frequently dealt with by well-known writers, this is, I believe, the first attempt made to bring together a body of evidence regarding the inoculability of leprosy and the evidence of its communicability by means of vaccination.

To physicians, superintendents of leper hospitals librarians of public institutions in Norway, Hawaii, the West Indies, the United States, and other countries, I am indebted for valuable official and other documents bearing on the subject, from which I have freely quoted.

The *British Medical Journal* of July 3, 1886, p. 24, in a leading article on the investigations into the causation of leprosy, undertaken by Dr. Edward Arning in Hawaii at the instance of the Hawaiian Government, observes that this inquiry "is likely to have results of great importance both to science and to practical medicine." This prophecy is indeed likely to be fulfilled, but hardly in the manner anticipated by the writer. Amongst the most important of Dr. Arning's discoveries is that vaccination has been instrumental in widely disseminating leprosy amongst the helpless and confiding population of that beautiful Archipelago.

Leprosy is one of the most loathsome as it is one of the most tissue-destructive diseases known, and when going through the wards of leper hospitals I have frequently noticed with pain the poor afflicted creatures bending their heads and covering their hands to conceal from strangers the sight of their distorted features and mutilated limbs. It is hardly possible to conceive, much less describe, the depth of human misery caused by the spread of this hideous and destructive disease ; but some idea of its nature may be gathered from the following description of leprosy, which may well excite the sympathy of the philanthropist. It will be found in a recent work on leprosy by Dr. Thin, pp. 99-100. It is translated from Leloir, an eminent French authority on leprosy, and refers to the tubercular variety of the disease. " If the patient," he remarks, " does not die of some internal disorder or special complication, the unhappy leper becomes a terrible object to look on. The deformed leonine face is covered with tubercles, ulcers, cicatrices, and crusts. His sunken, disfigured nose is reduced to a stump. His respiration is wheezing and difficult ; a sanious, stinking fluid, which thickens into crusts, pours from his nostrils. The nasal mucous membrane is completely covered with ulcerations. A part of the cartilaginous and bony framework is carious. The mouth, throat, and larynx are mutilated, deformed, and covered with ulcerated tubercles. The patient breathes with the greatest difficulty. He is threatened with frequent fits of suffocation, which interrupt his sleep. He has lost his voice, his eyes are destroyed, and not only his sight but his sense of smell and taste have completely gone. Of the five senses hearing alone is usually preserved. In consequence of the great alterations in the skin of the limbs, which are covered with ulcerated tubercles, crusts, and cicatrices, the pachydermic state of skin which gives the limbs the appearance of elephantiasis, and of the lesions of the peripheral nerves which are present at this time, and by which occasionally the symptoms of nerve leprosy are combined with those of tubercular leprosy, the sense of touch is abolished. The patient suffers excruciating pains in the limbs, and even in the face, whilst the ravages of the disease in his legs render walking difficult and even impossible. From the hypertrophied inguinal and cervical glands pus flows abundantly from fistulous openings. In certain cases the abdomen is increased in size on account of the liver, spleen, and mesenteric glands being involved. With these visceral lesions the appetite is irregular or lost. There are pains in the stomach, diarrhœa, bronchial pulmonary lesions, intermittent febrile attacks, and a hectic state. The peculiar smell, recalling that of the dissecting room, mixed with the odour of goose's feathers, or of a fresh corpse, is indicated, but badly described, by the authors of the Middle Ages, who compared it to that of a male goat."

Dr. John D. Hillis, formerly of British Guiana, says of the anæsthetic variety, that it is "known also as *leuke* of the Greeks, *baras* of the Arabians, joint-evil of the West Indies, *sunbahirii* of the East Indies, and dry leprosy, in contradistinction to the other form also known as humid leprosy; is characterised by a diseased condition of the nerves, and a peculiar eruption, the primary characteristic of which is loss of sensation, or anæsthesia; hence its name. After a time ulcerations form, a sort of dry gangrene of the limbs sets in, and joints drop off, and finally there is more or less paralysis. It would take a large volume to describe the signs or symptoms of leprosy, but the preceding account is sufficient to show what an alarming affection we have to deal with."

In Mrs. Hayes' little book, "My Leper Friends," is a chapter on leprosy, by Surgeon-Major G. G. Maclaren, M.D., in which the writer observes, pp. 123-4: "Acting on the strength of my own convictions as to the transmissibility and communicability of leprosy, I established the Dehra Dun Asylum on the principle already noted (the presence of a *bacillus* in the blood), and it has answered so far admirably; all its inmates living as happily as they can under their unfortunate conditions, and ending their existence contentedly! I have had, of course, ample opportunity of studying the nature of the disease, and its effects on the different organs of the body, and in many examinations I have made, *post-mortem*, I can testify that not a single organ in the whole body is exempt from the attacks and inroads of this dire and loathsome malady. It invades the brain, spinal nerves, the eyes, tongue, and throat, the lungs, the liver, and other digestive organs. In addition, as is generally known, it maims and deforms the external parts of the body in a manner too revolting to describe. It is painful to witness the amount of deplorable suffering some of these creatures endure. True it is that many feel but little pain—one of the forms of the disease producing *anæsthesia*, or insensibility of the parts affected; but this is the case in a few only. The majority suffer in variously painful degrees, according to the organ or part implicated, and it is a mistake to think that their sufferings are little. Many, in the earlier forms of the ailment, lose their sense of sight, smell, and taste, and when their lungs or throat is attacked (a common form), their agonies are dreadfully distressing and painful to behold. The inroads of the disease are slow and gradual, which makes it all the more trying, and the painful and lingering death to which most are doomed is a condition that one dreads to dwell on." It was the terrible nature of the disease that fostered the growth of the sumajh in India, the leper being accompanied to the grave with tom toms, where, in a sitting posture, he was buried alive.

In the West Indies, in British Guiana, in the Sandwich Islands, and in South Africa, when cases of invaccinated diseases were related to me, I was urged by the sufferers and by their friends to make known their grievances to English people and to the Imperial Parliament, and, if possible, to bring public opinion to bear upon a mistaken and mischievous system which, without doing the least good, has been the cause of such terrible and far-reaching consequences. Acting upon these entreaties, and upon others contained in communications from various leprous countries, I have presented to the public through the press, and to members of Parliament, such facts on this subject as came before my notice up to July, 1890. I now offer to the public further evidence and testimonies, on behalf especially of the afflicted population of our Crown Colonies and Dependencies, whose grievances have been so long and so flagrantly disregarded. Every attempt to introduce compulsory vaccination in the populous Island of Barbados, British West Indies, has been thwarted, owing to the widespread belief that leprosy and syphilis are communicated by the vaccine virus. In St. Thomas, Danish West Indies, and in Georgetown and other parts of British Guiana, it has, for similar reasons, been found practically impossible to enforce the vaccination law, and, in spite of severe compulsory enactments, entire districts remain unvaccinated by reason of this special danger; while, in the Sandwich Islands, a bill for the repeal of the vaccination law was introduced in the Legislative Assembly, July, 1890, by J. Kalua Kahookano, representative from North Kohala, Island of Hawaii.

Under the head of "The Legislature," the *Daily Commercial Advertiser*, Honolulu, November 9, 1892, publishes the recent report of the Sanitary Committee, as follows :—

"Hon. J. S. WALKER, President of the Legislature.

"SIR,—The Sanitary Committee report consideration of Bills Nos. 9, 13, and 25, and Petitions Nos. 33, 152, and 206, relative to vaccination.

"The object of all these bills and petitions is to repeal the law compelling parents and guardians to cause the children in their charge to be vaccinated.

"The complaints against the present system were very pronounced, and the repeal of the law making vaccination compulsory was strongly urged."

The committee met the Board of Health and conferred with them upon the subject, and under date of June 25th, addressed a communication to the Board, in which is the following:—

"Hawaiian Legislature, June 25, 1892.

"DAVID DAYTON, Esq., President, Board of Health.

"SIR,—An effort is being made in the Legislature to repeal or amend the law relating to vaccination; the object being to leave vaccination optional with parents and individuals.

"The chief objection raised against the present compulsory system appears to be the belief of some that leprosy, and other diseases, have been propagated by means of vaccination.

"It is said that some of the vaccinating officers are careless in the use of vaccinating instruments, operating first upon one person and then another without cleansing the instrument; and that there is distrust of the quality of virus used, in some cases serious inflammation and illness following the inoculation.

"The petitions and proposed measures relating to the subject have been referred to the Sanitary Committee, and the committee desire the views of the Board upon the subject. . . . . . .

"Any suggestions the Board may be pleased to make will be appreciated.

"Respectfully submitted,

"WILLIAM O. SMITH,

Chairman Sanitary Committee."

PREFACE.

The official report of the Honolulu Board of Health for 1892 shows that resistance to vaccination is spreading in many districts in these islands, and at the same time there is observed a sensible diminution in the number of lepers. In New Zealand, prosecutions for non-vaccination have for some time been abandoned. In the South African Colonies of Natal and Cape Colony the vaccination laws are enforced only during outbreaks of small-pox, and vaccination is everywhere regarded with mistrust. In the Transvaal and Orange Free State vaccination is entirely optional. In England there are about one hundred towns and poor law unions where the vaccination laws are a dead letter. In several of the Swiss cantons compulsory vaccination has been tried and abolished, and in no canton is there any penalty for non-vaccination. An attempt was made to pass a federal vaccination law in 1881, and was defeated in a Referendum by 253,968 votes against 67,820. In the Australasian Colony of Tasmania the compulsory law has been suspended by reason of its deleterious effects on the health of the people. In the Colonies of New South Wales and Queensland, Australia, the people have successfully resisted every attempt to impose the hotly-disputed Jennerian dogma upon them. Dr. Manning, the medical adviser to the Government of New South Wales, reports that in 1891 vaccination was only partially carried out in thirteen country districts. In ninety-two districts, no vaccinations were reported. The extent of the mischief already experienced will never be known, but sufficient is already admitted to arrest the attention of all who are seriously concerned for the public health and for the well-being of the community. Is it not, therefore, the duty of every medical practitioner to personally inquire into the matter for himself, and no longer to shelter himself behind the orthodox belief in the benign character of vaccination? For nearly a century Jenner's prescription has been tried and found wanting. Each of the reports of the Royal Commission on Vaccination already published establishes the failure, mischievous effects, and injustice of the compulsory infliction of an artificial disease upon healthy people, while some of the most distinguished names in the profession have testified to its being the certain vehicle for the dissemination of leprosy. These names include Sir Erasmus Wilson (sometimes called the father of dermatologists), Dr. John D. Hillis, Dr. Liveing, Sir Ranald Martin, Professor W. T. Gairdner, Dr. Tilbury Fox, Dr. Gavin Milroy; Dr. R. Hall Bakewell, formerly Physician to the Leper Asylum, Trinidad; Dr. A. S. Black, of Trinidad; Dr. Edward Arning; Dr. Walter M. Gibson, late President of the Honolulu Board of Health; Professor H. G. Piffard, New York; Dr. A. M. Brown, London; Dr. Frances Hoggan, Dr. Blanc, Professor of Dermatology, University of New Orleans; Dr. Bechtinger, of Rio; Professor Montgomery, of California;

Dr. Sidney Bourne Swift, late Medical Director, Leper Settlement, Molokai, Hawaii ; Dr. P. Hellat, St. Petersburgh ; Professeur Henri Leloir, Lille ; Dr. Mouritz ; Surgeon Brunt; Dr. John Freeland, Government Medical Officer, Antigua ; Dr. S. P. Impey, Superintendent, Leper Asylum, Robben Island, Cape Colony; and many others.

On the subject of leprosy there are no higher authorities ; therefore, considering that the evidence adduced in the following pages is founded upon an accumulation of facts and the testimony of eminent dermatologists, it is hardly open to doubt the intimate relation between the spread of leprosy and the increase of vaccination. May I not, then, urge that a concerted effort—by petitioning Parliaments, Legislative Councils, and other governing bodies, and by the powerful aid of the press—should be made to abolish the compulsory infliction of a disease fraught with such disastrous and far-reaching consequences to the human family? Until vaccination is disestablished and discontinued, and sanitary amelioration substituted for the inoculative experiments, drastic drug-medication, and nerve stretching, practised in various leper asylums, I am convinced that this dreaded disease will march onward with accelerated destructive force, and its ultimate extirpation will be rendered well nigh impossible.

No one can be more conscious of the shortcomings of this treatise than the author; but if the painful facts herein disclosed should induce some able and independent pathologist to continue these researches in the interest of the public and regardless of consequences, the author will feel that his efforts have not been altogether fruitless.

Rede Hall,
Burstow, by Horley, Surrey,
January 2nd, 1893.

THE

Recrudescence of Leprosy.

CHAPTER I.

THE INCREASE OF LEPROSY.

THE awakening of public interest in the leprosy question is due to the accumulation of evidence from nearly all parts of the world, showing that this fearful scourge, for reasons which are now being investigated, has greatly increased, and is still increasing.

At a dinner given at the Hotel Metropole, January, 1890, in aid of the National Leprosy Fund, at which the Prince of Wales presided, Sir Andrew Clark said that, "after making due allowances for the scare and disturbance which had been occasioned, there remained the obvious and indisputable fact that leprosy was a real question. He could produce overwhelming testimony of this fact, and the evidence was conclusive not only that leprosy did exist in larger measure in recent years, but that new germ centres were springing up in various quarters, and the old centres were widening. Before England and the civilised world there was looming a condition of affairs which might by growth threaten civilisation."

Sir Morell Mackenzie, in an article entitled "The Dreadful Revival of Leprosy," which appeared in the *Nineteenth Century* for December, 1889, after referring to its diffusion in Europe and America, says :—" In almost every other quarter of the globe leprosy is rife at present, and wherever it exists it seems to be slowly but surely extending its ravages. It is impossible to estimate even approximately the total number of lepers now dying by inches throughout the world, but it is certain they must be counted by millions. It cannot be comforting to the pride of England —'the august Mother of Nations'—to reflect that a very large portion of these wretched sufferers is to be found amongst her own subjects."

Dr. A. M. Brown, in his comments on "Leprosy in its Contagio-Syphilitic and Vaccinal Aspects," says, page 6 :—" From all that we can learn, leprosy is now alarmingly on the increase, particularly in some of our Colonial dependencies, and the fact has been causing much anxiety in later years."

The actual number of lepers throughout the world is far more than is stated in official statistics; for all authorities agree that there are cases, in some countries very numerous, which have never been reported, "the patient and his friends" (in the words of Dr. Charles W. Allen) " knowing with what horror the public regards the disease, naturally shunning publicity, and the physician humanely guarding his secret." From personal inquiries in many countries, particularly in Ceylon, Hawaii, South Africa, British Guiana, Venezuela, and the West Indies, I can fully confirm Dr. Allen's conclusion. In Hawaii leper-hunting is a dangerous business, as many of these unfortunate beings consider death preferable to the best managed lazaretto, where, besides loss of freedom and the companionship of fond relatives, they are obliged to dwell amidst the most repulsive and saddening surroundings. While on a visit to the General Hospital, Honolulu, my attention was called to a police officer, Kealioha Mani, who was lying severely, and probably fatally, wounded by a leper whom he was endeavouring to arrest. A short time ago a party of lepers armed themselves with five new Winchester rifles, and fired upon the police sent in pursuit. The love of freedom burns as brightly in these afflicted people as in their more fortunate countrymen.

In an article on leprosy in Hawaii, in the *Occidental Medical Times*, April, 1889, Dr. F. B. Sutliff, Sacramento, California, who spent four years as Government Physician at Wailuku, on the island of Maui, observes:— " The work of segregation has at no time been faithfully carried out. A large number of milder cases are not disturbed at all, and a good many others have been permitted to go free because of some influence, political or otherwise, that they may have possessed. . . . It is seldom that a leper desires to go away from his home to an hospital, and the study of his life after he knows himself to be a leper is how to live with his friends and keep out of the way of those whose business it is to know all about him. I never before saw a place where the people can hide so easily. . . . . They are quick to take alarm, and a look from the Government Physician, an inquiry concerning their name, is enough to cause them to change their residence at once." Dr. Sutliff says he has every reason to believe that there are at least four lepers not reported for every one that is.[1]

Dr. H. S. Orme, President of the State Board of Health, California, in his instructive treatise on Leprosy, says, "I have no doubt that the practice of *secreting lepers is general throughout the world*, wherever the disease prevails; and it is not difficult, in an early stage, for lepers to evade the authorities and go about their usual business."

The largest number of lepers segregated in any one year was in 1873, when the numbers received at the leper settlement, Molokai, according to the official reports, was 487, for several years previous to which arm-to-arm vaccination has been prosecuted with great and unparalleled energy.

The destructive results of this misguided policy are everywhere manifest for those who are not too prejudiced to see what is plainly before them.

RUSSIA.

Leprosy is reported to be increasing in Russia with startling rapidity. A St. Petersburg correspondent of the London *Standard*, January 18, 1891, says:—"The Town Council of Riga, aroused by the rapid spread of leprosy in the neighbourhood, has voted a sum of nearly six thousand roubles for the establishment of an asylum and hospital, which it is hoped will be ready to be

---

[1] Dr. L. Roussel, Government Medical Officer, in his report dated 30th October, 1888, Port Mathurin, Mauritius, says, "It is seldom lepers come to the Dispensary for treatment. Most of them hide themselves in the mountains, and do not like to move about in public."

opened in August. In 1887, Dr. Bergmann discovered thirty-seven cases in the town, and twenty-one in the environs. There are now over one hundred. In and around Dorpat, where the disease has attained alarming proportions, the late Professor Von Wahl strongly urged the necessity of leper colonies, such a system of compulsory isolation being in conformity with the provisions of the existing, but unenforced, law of Livonia."

As to the prevalence of "prokaza," or leprosy, in Russia, Dr. O. Petersen and Professor Münch have collected eight hundred and seventeen cases, which, however, must be considered far below the actual number. The former observer has noted forty-three cases from the records of the St. Petersburg hospitals in the last sixteen years.

Archdeacon Wright in his instructive work, " Leprosy an Imperial Danger," says that leprosy has increased so much of late in the Russian provinces of the Baltic that last year a "Society for Combating Leprosy" was founded at Dorpat, under the presidency of Professor Wahl, but otherwise composed entirely of lay members. Dr. Hellat, of Dorpat, travelled through the district, and showed that the reports made to the Government were very imperfect. In Livonia, where official statistics reported 108 cases, he found 276. In Courland he discovered 76 cases, and in Esthonia, 26. From other sources I hear that in the neighbourhood of Dorpat the lepers number as many as 17 per 1000, and another report says that in certain districts 10 per cent. of the population are affected. From more recent reports (May, 1892) I learn that the Town Council of Riga, alarmed by the ever-increasing proportions attained by the fearful malady, have just erected a leper hospital, at a cost of 60,000 roubles, which already contains 98 authenticated cases. The British Consul at Riga, in his report for 1891, says:—"It is difficult to discover the victims of this dire malady, as their relations and friends hide them from the sanitary inspectors."

In a communication from Dr. P. Hellat to Dr. P. S. Abraham on "Leprosy in the Baltic Provinces," dated 10th October, 1891, St. Petersburg, Mochovaia, 44, the writer says:—"My observations, continued for three consecutive years, gave the astonishing result that leprosy was very widely spread in the Baltic provinces; certainly considerably more than we formerly thought ourselves justified in believing. The number of lepers in certain districts is as much as 2 per cent. of the population. Furthermore, the investigation showed that the disease was steadily on the increase."—*Journal of the Leprosy Investigation Committee, No. 4, December, 1891, p. 7.*

BOKHARA.

The London *Daily Chronicle* of October 29th, 1891, contains the following (per Reuter's telegram):—

"St. Petersburg, October 27th.—The Emir of Bokhara has had several consultations of late with Russian medical men concerning the prevalence of leprosy in his dominions, and especially in the north-eastern quarter of the town of Bokhara, called Gonzari Pissiane, which spot may indeed be considered as the hotbed of leprosy in Central Asia. The lepers are allowed to lead an independent life in this quarter; they are allowed to contract marriages, and no supervision, whether medical or otherwise, is exercised over them. It has resulted from the advice tendered him at these consultations that the Emir has decided upon the foundation of a special hospital for the lepers, at which they will be treated by specialists in their disease."

NORWAY.

The present writer has been under the impression that leprosy had diminished in Norway, the diminution being generally admitted to be due to the segregation of lepers in the hospitals at Troudhjem, Molde, and Bergen ; but Dr. Vandyke Carter, who has closely investigated the subject, considers that:—"So far from leprosy in Norway showing a natural tendency to subside, there is ample evidence of a present activity equal to that displayed by the disease twenty-five years ago."—*British Medical Journal, Nov. 28th, 1885, p. 1048.*

ICELAND.

The Rev. W. T. M'Cormick, in a lecture delivered at Brighton, says:—" Before leaving, I was enabled to gain some details respecting leprosy, which is of a bad kind, and indigenous to the country, from Mr. Patterson, the British Consul, to whom Archdeacon Wright had written for information when publishing his book on this disease, and also Dr. Scheving. I learned here that in the year 1800 there were 150 cases out of a population of 50,000, but that now, out of a population of 72,000, the numbers had decreased to 25. I must state, however, that on further inquiries from an older and more experienced doctor near Laugardalir (Gudmunson, I think, was the name) I was told that the disease was increasing, and that one in every thousand was a sufferer from this hideous complaint. There are no hospitals for leprosy in Iceland, though Dr. Henderson, who travelled through the island in 1814, states that there were four then in existence.—*Journal of the Leprosy Investigation Committee, No. 4, December, 1891, pp. 69, 70.*

THE WEST INDIES.

During my visits to the Virgin Islands, the Leeward and Windward Islands, and British Guiana, 1888-89, I had opportunity of conversing with intelligent residents, including governors, medical practitioners, superintendents of leper hospitals, magistrates, prison chaplains, editors of newspapers; and the general opinion was that leprosy was largely on the increase. In some islands, such as Jamaica, St. Kitts, and Trinidad, there are leper communities, which are gradually increasing; and appeals are frequently made in the Colonial Press for their segregation in hospitals.

In a dispatch to the Colonial Secretary, Dr. R. Hall Bakewell, Vaccinator-General, Trinidad, said:—"The very great increase of leprosy in this island, particularly among persons in easy circumstances, is the subject of general remark, and although we have no statistical evidence of the fact, yet it seems admitted on all hands."—*Page 7, Compendium of "Extracts from Report and Returns" in the "Royal Gazette," Trinidad, March 1st, 1871.*

On the 22nd January, 1889, I visited the lazaretto at Barbados, a crowded institution. A new ward was then in course of construction, to accommodate 32 more patients; but the applications from the single parish of St. Michael were greater than the extra beds to be provided. I may mention that the island of Barbados comprises thirteen parishes, with a total population of about 180,000, of which St. Michael's contains about a sixth; and it is estimated that 150 to 200 more beds ought to be provided under the present system of voluntary segregation. If the segregation, which includes only the leprous poor and pauper class, were compulsory, as some now demand, the alarming spread of the disease, which is endemic in all the islands, would be yet more fully exhibited.

The *Official Gazette*, Barbados, May 5th, 1890, p. 524, says:—"With a daily average of 104 there have been 16 admissions, 3 discharges, and 4 deaths. The Poor Law Inspector, Mr. C. Hutson, says:—"Considering the overcrowding of the wards, it is, I think, wonderful that we keep so clean."

The census returns from Barbados show that while the population during the ten years, 1871-81, had only augmented 6 per cent., the lepers had multiplied 25 per cent.

According to the Surgeon-General's report of hospitals in Trinidad for 1880, No. 41, p. 38, the number of patients in the Leper Asylum on June 30, 1880, was 124.

The report for 1888, *Trinidad Royal Gazette*, p. 1116, says:—"The admissions have been limited to the amount of accommodation, and there were fourteen lepers at the end of the year in the Colonial hospitals, awaiting vacancies for admission to the Asylum."

The Asylum (at Mucurapo, Port of Spain, Trinidad), which I visited in February, 1889, contained at that time

180 patients (under the medical superintendence of Dr. Bevan Rake), who are admirably cared for by the French Dominican Sisters. Every bed is occupied. In his report to the Surgeon-General for 1887, Dr. Rake says: —"The new infirmary at the Asylum was opened in August last, and was quickly filled, 19 patients being admitted on the 19th, and nine more on the 25th. Since then it has been constantly full."

I was informed by the lady superintendent that a new ward was to be built at once, to contain 30 additional beds. There were then, she said, fourteen lepers in the Colonial (Port of Spain) Hospital awaiting vacancies for admission to the asylum.

In the last report on leprosy in Trinidad, dated March 1st, 1891, by Dr. W. V. M. Koch, acting Medical Superintendent, it is stated (p. 65) that the new infirmary ward, which was finished at the end of 1889, and occupied early in 1890, has been full all the year round. There was a rush of patients to fill it.

The *Trinidad Leprosy Report* for 1890 (p. 31) says that during the year a new ward containing 30 beds has been opened. The asylum contains 210 inmates, "every bed being occupied."

Dr. Bevan Rake says :—" There is, I fear, little doubt that the disease is increasing in Trinidad as in other tropical countries."—*Papers on Leprosy, Trinidad, p. 34.*

In an article entitled the "Dreadful Revival of Leprosy," in the *San Fernando Gazette*, Isle of Trinidad, 22nd February, 1890, the writer says :—" It may not be generally known that as far back as 1805 there were only three lepers in Trinidad ; eight years later there were 73 out of a population of 32,000. Twelve years later, when an attempt was made to segregate them upon a small adjacent island, it was found that these afflicted persons had increased so rapidly that the scheme had to be abandoned. In 1878 there were 860 out of a population of 120,000, and later statistics show that the number of lepers was increasing four times as rapidly as the population." The writer arraigns the authorities for their supineness, and urgently calls upon them to take the necessary steps to arrest the progress of this fearful disease.

In a leading article in the *St. Christopher Gazette* (of St. Kitts), the 17th May, 1889, entitled, " The most pressing question in the Colony," the writer quotes Dr. Boon's last quarterly report, which (he says) "clearly and forcibly showed the Government the enormous increase in our leper population during the last six years." Dr. Boon, who held the position of Acting Government Analyser of Vital Statistics, says : — " There is one subject to which I would specially call the attention of the Government, and that is the necessity of legislation with regard to lepers. I am satisfied that the disease is increasing rapidly in this island (St. Kitts)."

In the *Lazaretto*, No. 11, a paper published in the West Indies, the editor asserts that a careful census carried out by medical officers would demonstrate that St. Kitts and Nevis contain more lepers per thousand of the population than any other British possession. He also considers that the disease has increased in Antigua, and there are no fewer than 300 lepers in the Leeward Islands. The *Lazaretto*, No. 21, for April 20th, 1891, estimates the number of lepers in the two islands of St. Kitts and Nevis at 200. In 1871 Dr. Munro discovered, by a personal census, that there were 72 lepers at St. Kitts, a number which has now increased to 135, or at the rate of 90 per cent. in twenty years. To accommodate the growing community of lepers, a large lazaretto has recently been built at Sandy Point, ten miles from Basse Terre, St. Kitts, which already contains eighty inmates. The *British Medical Journal* for June 20th, 1891, says that a petition was lately sent to the Governor, Sir W. F. Haynes-Smith, with the request that it might be sent to the Queen. Amongst other things it states that leprosy is most prevalent in these islands, and that the number of persons afflicted with it is rapidly increasing.

The London *Daily Graphic* for August 15th, 1891, publishes the following:—" Sir Morell Mackenzie writes —' I beg to enclose a copy of a letter recently received by Dr. Munro, formerly medical officer of St. Kitts, West Indies, which shows conclusively that leprosy is extending in that colony.'

" The letter from Dr. Boon, referred to by Sir Morell, runs as follows:—' In your time I believe there were about fifty lepers in St. Kitts ; at present there are 120 known lepers, and I think there are a good many more that are kept hidden from the medical men. I am at present getting as complete a list as possible of the lepers here. One thing is very noticeable in Nevis, namely, the way in which the leprosy spreads in each neighbourhood from single cases. It is not easily traced in St. Kitts, as the people there do not own land like the Nevis people, and are consequently more nomadic. One thing has struck me very much, and that is the number of shop-keepers that have contracted the disease.'

"We have also received a letter from Dr. Boon, who says:—'The enclosed photograph of mendicant lepers, subscribed for by a few gentlemen of this island for the purpose of forwarding to you for publication in the *Daily Graphic*, will give a slight idea of the risks by contagion to which the population of this colony is daily subjected. Leprosy has attacked people of all conditions in the West Indies. A few years ago a newly-appointed inspector of police enforced the local 'Vagrant Act,' and prevented the squads of mendicant lepers from perambulating the town, begging from house to house, and importuning people in the streets. Through the action of the then President of the Island, the inspector was forbidden by the Governor to interfere in any way with these lepers. The fact that a member of the former gentleman's family was afflicted with this disease may have had something to do with his action in the matter. We count among our lepers (other than mendicants) bakers, butchers, salesmen in groceries and provision shops, fishermen, printers, editors, circulating-library keepers, shopkeepers, planters, agricultural labourers, and carpenters. In a lodging-house kept by a leper, members of the Bar lodged when on circuit, and slept on the same bed used by the leper when he had no lodgers. Another leper kept a crèche, and tended about twenty infants at a time in his room for over ten years.'"

In the report of the Blue Book of St. Vincent, British West Indies, 1890, the Acting Administrator observes:—
"It is greatly feared that leprosy, which has already proved so great a scourge to some of our colonial possessions, will become a serious trouble in St. Vincent.

Our administrator of the islands, Mr. Irwin C. Maling, reports to the Colonial Office that the disease, though perhaps slowly, is surely on the increase; and though the average of patients treated at the Leper Asylum is only 15, there are many more at large. No law exists to compel those afflicted with the disease to go to the asylum and receive proper medical attention, but the subject is one which will receive the early attention of the Local Government."

On June 2nd, 1890, Mr. Gourley, M.P., called the attention of the Under-Secretary of State for the Colonies to the considerable population of lepers in the West Indies, which, he said, was daily increasing. Sir W. F. Haynes Smith, of the Leeward Islands, who informed me in 1889 that leprosy was seriously on the increase in the West Indies, issued an address to the Federal Council in April, 1891, in which he quotes the opinion of the medical profession that the disease has greatly increased, and that the only satisfactory explanation of the spread of the disease is that under certain conditions it is communicable and contagious. Sir William Robinson, Governor of Trinidad, writing to the Secretary of State from Government House, 9th May, 1889, says:—" After fifteen years' residence in the West Indies, I can fully corroborate Dr. Rake's statement that leprosy is on the increase, and I am not surprised at it."—*Papers on Leprosy, Trinidad, p. 35.*

In the report of the superintending medical officer of Jamaica, dated July, 1891, it is stated that 420 cases of leprosy in the island are known to the district medical officers, but it is conjectured that a good many living in unfrequented districts are not reported, and that some cases in families of the better classes are not to be found in these returns. The actual number is estimated at 450, or one in 1378 of the present population of the island, reckoned at 620,000. This is a decidedly low estimate. A new ward, capable of accommodating forty beds, has just been made ready for the reception of these unfortunate patients.

A communication from Dr. N. Lacary, physician to the lepers in the French Antilles, dated Basse-Terre, Guadaloupe, January 16th, 1892, and sent by request of the chief of the Sanitary Department at Guadaloupe, in reply to an inquiry for the statistics in respect to leprosy, states that it is impossible to report with accuracy the number of persons known to be more or less subject to leprosy throughout the various districts. The island of La Disirade, in which the Lazar-house is situated, may afford some exact figures, and contains one hundred diseased (leprous) persons from Guadaloupe and its dependencies, and from Martinique. The lepers frequently secrete themselves, and it is impossible to give the exact figures of those who are at large. It is recognised throughout the islands that leprosy is on the increase.

JAPAN.

As my communications to the Authorities in Japan requesting information remain unanswered, I have but few details to report. Leprosy has existed from time immemorial in this country, and there is an old established leper settlement at the hot springs at Kusatsu, to the north of Tokio. Leprosy is reported to be increasing considerably, and according to a communication in the *Liverpool Mercury*, September 22nd, 1891, three leper hospital asylums have been established in Tokio during the past ten years. One of these hospitals treated in 6 years 4249 cases, of which 3852 were those of Japanese patients. The *Pioneer* (Allahabad) of September 9th, 1891, reports that leprosy has spread in the Japanese villages to an alarming extent during the past few years. In one village near Toimachi, in the Gifu Ken province, every inhabitant is a leper. The Japanese Government is taking steps to look after their afflicted people.

The Lancet for May 25th, 1889, states that vaccination was made compulsory in the seventh year of Meiji (1874). It will be noted therefore that the rapid diffusion of leprosy took place shortly after the introduction of the compulsory law, and has kept pace with the progress of vaccination in this community. Dr. Tamanoto, of the Imperial Japanese Navy, says that when leprosy occurs in a family it is the custom to conceal it.

The Rev. Father Vigroux, Missionary Apostolic, in an article in the *Catholic Review*, which also appears in the *Tablet*, May 14th, 1892, says that 44 patients have already been admitted to the Leper Hospital at Gotemba, Japan, founded by the late Father Testevuide. There is accommodation for as many as eighty.

THE GRECIAN ARCHIPELAGO, TURKEY, AND SYRIA.

In the Island of Samos, with a population of 42,000, there are 43 registered lepers, and many others unregistered ; to segregate whom the Prince of Samos has built an asylum. In Constantinople Dr. Dujardin Bourmetz estimates that there are three thousand, some of them at large, and others in the hospitals.

In Syria lepers abound, and the most repulsive examples I ever saw were in the lazaretto at Damascus, where the supervision and accommodation was of the most wretched description. For their comforts they depended on the alms given by casual visitors. In Palestine I noticed many lepers, in the most hideous state of deformity, begging. This was in 1884. The Constantinople correspondent of the *Times* writes (July 31st, 1889):—" Dr. Zambaco has made a special study of leprosy, and purposes to present to the congress the result of his assiduous labours on the subject. Like some others, Dr. Zambaco has come to the conviction that leprosy is non-contagious. He offers practical arguments and proofs in support of his opinion. There are in Constantinople alone upwards of 250 lepers, all of whom Dr. Zambaco has personally attended. Of these, 25 individuals only are isolated in a special locality at Scutari—the remainder are to be seen in the streets, and in contact, without any restraint, with the rest of the population. In the Islands of the Archipelago there are at Crete 8000 lepers ; and at Rhodes, Cyprus, Mytilene, Tenedos, and other smaller islands, they are also numerous, and excepting in the larger ones, free in their movements. Dr. Zambaco has prepared to lay before the congress, and for publication, a most interesting work upon the subject of leprosy in the Levant, containing numerous illustrations, portraits, and biographies of patients living and dead, with accounts of curious cases of cure, non-contagion, and remarkable facts observed by him, which cannot fail to attract the earnest attention of scientific men with respect to the disease which has of late come so prominently before the public."—*Times, August 6, 1889, p. 10.*

EGYPT.

During my visit to Egypt in February, 1891, I endeavoured to ascertain the facts as regards leprosy in that country. I called at several hospitals, and conferred with a number of officials connected with the Government, but without obtaining much information. Dr. Selim Zeidan, the medical director of the new General Hospital at Luxor, informed me that during the past few months five lepers had presented themselves for admission, but he was obliged to refuse them. This was due not to want of space, but to lack of funds. In a communication to Dr. Abraham, dated 4th November, 1890, and published in the *Journal of the Leprosy Investigation Committee*, Dr. Greene, of Cairo, director of the Sanitary Department, Egypt, states that the total number of cases reported is 2058, "but this does not by any means represent the whole of the lepers in Egypt, for many districts, where I have reason to suppose some exist, sent in blank returns."

THE UNITED STATES.

Dr. Blanc, of New Orleans, Clinical Lecturer on Dermatology at the Medical College, and Dermatologist to the New Hospitals, has had special opportunities for observation. In a report to the

Louisiana Board of Health, May, 1889, he refers to 42 known cases of leprosy at New Orleans, twelve cases at La Fourcho, and six cases (these doubtful), at St. Martinsville; and in only a very small number could he discover the causation in heredity.

In the *New York Medical Journal*, July 13th, 1889, Dr. R. W. Taylor states that during the past fifteen years he has almost constantly seen from one to three lepers in the crowded wards of the hospitals on Blackwell's Island, New York. Other authorities give similar reports of the New York hospitals.

Dr. Charles W. Allen, Dermatologist, in his article on "Leprosy" in the *New York Medical Journal*, March 24th, 1888, calculates the number of lepers in the United States at 150. "Unquestionably," says Dr. Prince A. Morrow, "is leprosy gaining ground in this country, and the disease prevails over more than a fourth of the habitable globe."

THE SANDWICH ISLANDS.

In the Sandwich Islands leprosy is allowed to be the chief of the destructive forces which are gradually depopulating the native race of this beautiful archipelago. Its rapid increase is by far the most urgent and anxious question of the hour, and successive Medical Officers of Health seem powerless to cope with it.

In a leading article on "The Nature of Leprosy" *The Lancet*, July 30th, 1881, p. 186, says:—"The great importance of the subject of the nature and mode of extension of leprosy is evident from the steady increase in certain countries into which it has been introduced.

In the Sandwich Islands, for instance, the disease was unknown forty years ago, and now a tenth part of the inhabitants are lepers. In Honolulu, at one time quite free, there are not less than two hundred and fifty cases; and in the United States the number is steadily increasing."

According to the latest returns handed to me (October, 1890) by Mr. Potter, the Secretary to the Board of Health, Honolulu, 1154 lepers were segregated in Molokai, to which must be added thirty, sent from the Hospital of Suspects at Kalihi to Molokai on the 30th of the same month, while there are probably several hundred secreted by relatives in the various islands. On 31st March, 1888, the number officially reported to be at large in the various islands amounted to 644, but efforts have been made during the past three years to capture these afflicted creatures and segregate them at Molokai.

The experiment of segregation, coupled as it is with enforced separation of relatives and friends amongst a race who are very gregarious and affectionate, is attended with great difficulties, and found impossible to carry out successfully.

A medical practitioner informed me that well-to-do lepers were not interfered with, and he gave me the names of several at large who occupied prominent positions in the island; and he observed that it was not intended to disturb them, notwithstanding the law which imposes segregation upon all lepers regardless of distinction.

Dr. J. H. Kimiball, of Honolulu, ex-President of the

Board of Health, in his official report for 1890, observes: —"Some very bad and unmistakable cases are in hiding in the fastnesses of the mountains or high up in the valleys, fed and secreted by their friends, while some mild cases change their place of residence so often as to baffle the efforts of the officers of the law for their arrest."

Dr. Edward Arning reported that he had visited the remotest gulches and corners of the islands, where few white men penetrate, and had found lepers at large, including some bad cases. He suggests that it may be just as well to leave these poor wretched creatures where they are, as they are more out of the way there than at Kakuako or Kaluapapa. Dr. John S. M'Grew, in a communication to General James de Comby, the United States Resident Minister, says:—"From political and other influences with officials of the Government, many lepers are permitted (in Hawaii) to go at large without being questioned—really dangerous cases of leprosy."

Dr. White, surgeon to the United States Navy, who visited the islands in 1882, in a report to his Government estimates the concealed cases at 3 per cent. of the population.

Mr. Dayton, the President of the Board of Health, Honolulu (an old resident in the island, who has had a wide experience in the service of the public health), was kind enough to furnish me with facts relating to the introduction, establishment, and increase of leprosy throughout Hawaii, and the steps taken to deal with it by isolation, medical treatment, and hygiene, and also with copies of official reports published by the Board of Health.

According to the same writer on the subject, leprosy was discovered in the island in 1840, but Mr. D. W. Meyer, Agent for the Honolulu Board of Health, in the appendix to the report presented to the Legislative Assembly of Honolulu in 1886, says it was in 1859 or 1860 that he saw the first case of the disease. That 1840 was the date of its introduction is the opinion of Dr. W. B. Emerson, ex-President of the Board of Health, Honolulu, who, in his report published in *The Practitioner* of April, 1890, attributes the introduction of the disease to a case reported by the Rev. D. D. Baldwin, M.D., to the Minister of the Interior, May 26th, 1864. In 1863 Dr. Baldwin received reports from the deacons of his church at Lahaina with the names of 60 people who were believed to be affected with this disease. In a very few years leprosy increased to an enormous extent, and in 1868 Dr. Hutchinson reported 274 cases.

Dr. Emerson says:—" Leprosy has made fearful strides. It is not necessary to trace with precision the curve that represents the increase of leprosy in these islands from that date to the present time. It is a fearful story, and should teach us that leprosy is undoubtedly communicable."

In his report dated Molokai, March 31st, 1888, Mr. Meyer says:—" That the spread of this scourge in these islands has been truly fearful is known to every one here, and that it could not have spread as it has done unless it were communicable, appears to me to admit of no doubt."

To account for the appalling spread of this terrible scourge of humanity within such a short period of time,

INCREASE CONCURRENT WITH VACCINATION.

the evidence points conclusively to one prominent cause —vaccination. There is no evidence to show that leprosy increased in Hawaii until after the introduction and dissemination of the vaccine virus.

Small-pox was introduced from San Francisco in the year 1868. In that year a general vaccination took place, spring lancets being used, which the President of the Board of Health (Mr. David Dayton) informed me were difficult, if not impossible, to disinfect—the operation causing irreparable mischief. The synchronicity of the spread of leprosy with general vaccination is a matter beyond discussion, and this terrible disease soon afterwards obtained such a foothold amongst the Hawaiians that the Government made a first attempt to control it by means of segregation. Another outbreak of small-pox occurred in 1873, and yet another in 1881, both followed by general arm-to-arm vaccination and a rapid and alarming development of leprosy, as may be seen in successive reports of the Board of Health. In 1886 the then President of the Board of Health recorded his conviction, in an official report, to the effect that "to judge by the number of cases in proportion to the population, the disease (leprosy) appears to be more virulent and malignant in the Hawaiian Archipelago than elsewhere on the globe." Leprosy became then, and is now, the most pressing question in these islands.

VISIT TO KALIHI.

In October, 1890, I started from Honolulu in company with Mr. C. B. Reynolds, the Chief Executive Officer of the Board of Health, to visit Kalihi, a place three miles away, where persons supposed to be tainted with leprosy were incarcerated. It consists of cottages, dispensary and recreation ground, the whole surrounded by a double wall to prevent escape, and is a dreary place of abode. There were 74 patients in the establishment, most of whom exhibited distinct traces of this loathsome malady, including some unusually bad cases. Within this enclosure was a comfortably furnished cottage which Sister Rose Gertrude had recently occupied, but had now vacated. Both the cottage and the dispensary of which this lady had the charge were in a bad state of disorder, and presented a painful contrast to rooms in the care of the Dominican Sisters at Trinidad, and in other Asylums I have visited.

The day after my visit, 31 of the inmates were taken from the Suspect Hospital to Honolulu, and thence from the King's Wharf were forcibly deported to the living grave at Molokai, from whence no traveller returns. The scene was of the most painful description. These afflicted creatures were torn from their friends and relatives, husbands from wives, children from their parents, frantic with uncontrollable grief. Lovers were separated, their lips trembling with emotion, amidst unutterable wailings, wringing of hands in the agony of despair, and heart-breaking experiences which I shall never forget, and which the pen of a Balzac, or Victor Hugo, could only adequately describe. Mr. Reynolds said that at times it seemed more than he could stand and he did all that was possible to mitigate their sufferings. He told me that in ten days' time there would be another contingent to undergo the same sorrowful experience. This goes on year after year, and will probably continue until medical men themselves turn their attention from experimental treatment to preventive measures, and themselves petition Governments to suppress the mistaken system of vaccination, which, it is admitted by the highest authorities, has been a prolific source of this terrible evil.

THE MAURITIUS.

In the Mauritius, according to Dr. Seizor, in the *Progrès Medical*, 1886, translated by Dr. P. Abraham, and quoted in his work on leprosy, lepers of all races, including Europeans, can be counted by hundreds without difficulty.

MADAGASCAR.

The Paris correspondent of the *Daily News* telegraphs, August 12th, 1890, that the French Catholic missions in Madagascar have taken up the lepers there, and have built a lazar-house at Ilafy, near Antananarivo. A second hospital for 200 lepers, who are not so invalided, is at Abohivaraka, where they work in the rice marshes.

CANADA.

In May, 1885, the lazaretto at Tracadie, New Brunswick, contained 21 lepers, and others were known to be at large.

The *British Medical Journal*, July 18th, 1891, says:— " Owing to the increase of leprosy in British Columbia, the inhabitants recently memorialised the Canadian Government, asking that some steps should be taken to check the progress of the evil. It has accordingly been resolved to found a leper colony in D'Arcy's Island, which lies off the coast. As soon as the arrangements are complete all the lepers in the colony (most of whom are Chinamen) will be transferred to this island. Dr. F. H. Smith, the superintendent of the well-known lazaretto at Tracadie, New Brunswick, has been requested to investigate the alleged increase of leprosy in the towns of the Pacific slope of the Dominion."

UNITED STATES OF COLOMBIA.

In the consular report to the Foreign Office, sent by Mr. T. H. Wheeler, the British Consul at Bogotá, dated September 26th, 1890, I find various facts confirming previous reports as to the rapid development of this disease. Referring to the number of lepers, Mr. Wheeler says:—" A well-known physician of Bogotá, the editor of the *Medical Review* of the city, has stated, in an article lately published in that review, that one-tenth part of the inhabitants of Santander and Boyacá are lepers. As the population of these two departments is about 1,000,000, this estimate would give 100,000 lepers in that portion of Colombia alone. The medical officer of the principal lazaretto of the country, who has travelled extensively in the departments of Santander and Boyacá, with a special view of studying the question of leprosy, estimates that they contain some 30,000 lepers. I believe that both these estimates are very much exaggerated, but the number of lepers must be very great in those two departments, as I am aware from my own observations in travelling through them in 1883. Some months ago the Government of Santander endeavoured to obtain from the various municipal authorities official returns of the number of lepers in the department. Reports were sent in from about half the municipalities, giving notice of 1804 lepers, of whom 388 were living in the lazaretto of the department. But no returns were sent in from the districts known to be most infected with leprosy, and in any case, no satisfactory census upon such a subject can possibly be taken in Colombia. On the merest suspicion that anything of the sort was intended, every leper, who could possibly do so, would at once disappear, and remain hidden until it was all over, for fear of being dragged to a lazaretto and separated from his family and friends, and amongst the population, in general, there is no such dread of the disease as would lead them to give information against any lepers who might be concealed in their neighbourhood.

"Santander and Boyacá are the departments most infected with leprosy ; next to these are Cundinamarca, Tolima, and Antioquia, in the order named, but a certain number of cases are to be found in all the other departments. In Cundinamarca, where the number is more easily estimated than in any other part of the country, there are said to be over 4000 cases of leprosy, and in Antioquia from 800 to 1000. In the whole Republic, there cannot, I believe, be less than 20,000 cases, and there are probably many more.

"Leprosy is most common in Colombia amongst the lowest classes, whose homes and mode of life render them liable to exposure to cold and damp, but it is by no means confined to the poorer class, as many people of wealth and position have been attacked by it."

There are two lazarettos in Colombia, one in Agua de Dios, where one consul found 520 lepers in July, 1890, the other at Contratacion, department of Santander, with 390 of these afflicted people. A Bill has recently been laid before Congress asking for an appropriation of £10,000 for the construction of a new lazaretto, but the disease has reached such proportions as to render such a sum quite inadequate to cope with the ravages of the disease. Mr. Wheeler says that it seems inevitable that leprosy should continue to spread more and more throughout the country, unless some great effort is made to arrest its progress. He remarks that in Antioquia not a single case of leprosy was known thirty years ago. Since then, the disease has spread in all directions, and the number in this town is now said to be over 800. I may add that, during the interval, vaccination has been introduced in all the Republics of South America with the usual sinister results.

Mr. Edward H. Hicks, M.R.C.S., practitioner in Bogotá, says :—" The local authorities have called attention to the alarming spread of leprosy in the Republic of Colombia, South America. In districts in which the disease was formerly unknown it has appeared to spread with great rapidity. As to articles of diet, the greater number of cases occur where fish cannot be obtained in these places."—*British Medical Journal, Nov. 8th, 1890.*

The *New York Herald*, of April 10th, 1892, says:— " Reports about the spread of leprosy in the Republic of Colombia are of the most alarming character, and should receive the very serious consideration of the United States health authorities. Every district in Colombia is said to be more or less infected."

In the last annual report of the Consul-General of the United States of Colombia (1890-1) it is observed that the question of the great increase of leprosy has become a very grave one for the country. Prior to the year 1860, the numbers of the afflicted were comparatively stationary; but since that time the increase has been much more rapid, and the disease has spread to districts in which it was previously unknown, almost the whole of Colombia being now infected. Children are frequently seen in the lazarettos in a leprous state, and the mortality is exceedingly high.

Dr. Alzevedo Lima, chief medical officer, Hospital, Rio de Janeiro, says :—" Has leprosy increased in Brazil within the last years? It seems that it has, but we have no exact data to guarantee a positive affirmation. However that may be, it is no exaggeration to say that in Rio de Janeiro there are more than 300 lepers disseminated throughout the city."—*Journal of the Leprosy Investigation Committee, December, 1891, p. 25.*

The last census of Brazil returned the number of lepers at 5000, but Dr. Lutz, a lepra specialist, estimates it at 10,000 and upwards.

BRITISH GUIANA.

In January, 1889, I visited Demerara and Essequebo, British Guiana, but owing to the state of my health I was unable to visit the leper asylums at Mahaica and Gorchum. During my sojourn, the newspaper at Berbice (where the leper asylum is located) published a statement to the effect that, for want of accommodation in the asylum, lepers in the worst stages of the malady had been seen in the streets and bye-ways. In the report of the Surgeon-General of British Guiana for 1887, Dr. C. F. Castor says : — " I hear on all hands that leprosy is spreading — not only here, but all over the world—

very considerably." And in a communication to the secretary of the Leprosy Committee, dated 21st November, 1891 *(Journal No. 4, p. 36)*, Dr. Castor observes :—" No one for a moment doubts that leprosy is spreading." I was able to obtain only one report of a date earlier than 1887, that of the Surgeon-General for 1879, in which Dr. Manguet says :—" Many young children are brought to me in the incipient stages of the disease (leprosy)," and added that the disease was spreading.

When going over the Colonial Hospital, Georgetown, British Guiana, Dr. Ferguson spoke to me of the serious increase of leprosy in the colony, and said that they were obliged at the general hospital to receive lepers for whom there was no room at the asylum. He pointed out to me five lepers in one ward with other patients. Dr. J. L. Veendam, a Government medical officer, who had at that time resided sixteen years in the colony, and has medical charge of four sugar estates, assured me that leprosy was much more widely disseminated in the colony than was generally supposed, and that this was the case amongst all classes of society. Referring to a ball which was to be given that evening to Governor Haynes Smith, he added, " I know lepers who will mix with the gay throng this evening." Some time ago Dr. Veendam medically examined all the labourers on the four estates under his charge, 250 in number, and found about 50 who were more or less tainted with leprosy.

One of the highest authorities, Dr. John D. Hillis, F.R.C.S., for ten years Medical Superintendent of the principal lazaretto in British Guiana, and who has devoted twenty years to the consideration of this important subject, says :—"To the most casual observer (in British Guiana) the increase must be apparent, irrespective of the fact that the asylums cannot be enlarged fast enough to contain the cases that are compelled, by want or the rapid advance of the fell disease, to seek admission and relief within their walls, whilst hundreds of others, it is well known, do not enter but remain outside to mingle with others or contaminate their surroundings. Not only is leprosy on the increase in the colony, but the increase has been greatest in the last decade. . . . Wherever the writer goes, he meets with lepers walking about among, and mixing with, the people, may be in the church, or in shops. As the signs and symptoms of the disease become better known, they will perhaps be more easily recognised by the uninitiated."

In 1858, the lepers were located at the present Institution at the mouth of the Mahaica Creek, which not very long ago was enlarged to meet the ever increasing demand on its accommodation.

On 31st December, 1859, there were only 105 inmates at the Asylum. In 1869, they had increased to 300, and the place could hold no more. Increased space was provided, and in 1889 we find, from the official reports, over 500 of these unfortunate inmates. Around this leper asylum, outside its boundaries, there are large numbers of lepers not included in these returns. Dr. Hillis states that, while the increase of the population in twenty years, between 1858 and 1878, was only 45 per cent., the increase of lepers was 160 per cent.

BRITISH GUIANA.

Dr. George Thin, in his work on "Leprosy," p. 76 (1891), says :—" It is estimated that in 1890 the number of lepers in British Guiana was 1000, or one in every 250 of the population." My own inquiries led me to suppose the number was larger, as from all that I could gather there were about 500 lepers in the two asylums at Mahaica and Gorchum, and I saw several at the General Hospital, Georgetown, for whom there was no room in the asylum. These were reckoned by old medical residents to be far less than half the total leper population.

DUTCH GUIANA.

Leprosy is making rapid progress in Dutch Guiana, and a devoted priest, who has been attending to the temporal and spiritual wants of the people, was reported in October, 1890, to be dying of the disease. Bishop Walfingh, of Surinam, has recently visited Holland, with the object of raising funds to build a suitable hospital, and has met with a successful response to his benevolent appeal.

VENEZUELA.

In the early part of 1889 I visited Venezuela, pursuing my inquiries as to leprosy in Carracas, Bolivar, and other places. From all the information I could obtain I learned that its spread, though less conspicuous than in the adjoining territory of British Guiana, is making constant advance. From a report made by the United States Consul at Maracaibo, Mr. F. H. Plumacher, to the American Government in 1890, I find that leprosy began to be felt as early as 1828, and in 1841 the National Government, under the direction of President Bolivar, purchased an island four miles east of Maracaibo, and erected an hospital and dwelling-houses for the accommodation of these afflicted people. In 1876 the cases had assumed alarming numbers, so as to seriously endanger the sanitary future of the State. In the year 1890 there were 125 patients in the lazaretto, and many more at large in the city and environs, and all attempt to segregate them is thwarted by the efforts made by friends for their concealment. This increase here, as in other countries, is coincident with the extension of the practice of vaccination.

AUSTRALASIA.

Leprosy is not unknown in the Australasian colonies, and is especially noticed in a report dated 7th May, 1890, and ordered to be printed by the Legislative Assembly of New South Wales. This report was handed to me by the President of the Board of Health. From it I find that, at the close of 1889, there were 30 cases of leprosy under official cognizance.

The *Lancet*, of August 1st, 1891, says that the number of lepers has more than doubled during the past ten years.

Referring to the report for 1891 (which has not yet reached me) the *Sydney Mail* of February 20th, 1892, under the head of "Leprosy," observes :—" Those of the public who are the least disposed to alarmist views will probably regard as highly serious the statements now made public regarding leprosy in New South Wales. The statements give a great shock to the feeling of confidence based on the supposed comparative immunity of persons of European races from the attacks of this terrible disease. Of course, it has always been known that where the conditions are specially favourable to contagion, the supposed racial protection ceases to be a safeguard. But it is, nevertheless, startling to learn that of ten persons found by the Board of Health during the year to be suffering from leprosy five were natives of this colony of European descent, while four were Chinese, and one a Kanaka. At the beginning of the year there were 13 lepers under detention at the lazaretto, while those so detained at the end of the year numbered 21, of whom eight were natives of New South Wales of European descent, 11 Chinese, one Javanese, and one Kanaka. During the time the lazaret has been available there have been 31 patients, of whom one could not be detained, and nine have died."

Our colonists are becoming alarmed at the invasion of leprosy in New Zealand. The *New Zealand Herald*, Auckland, June 14th, 1890, in an article headed " Leprosy Among the Natives," says :—" A gentleman at Hokianga writes to a friend here—' My brother was north a few days ago on a vaccination tour at Herekino, and he reports an outbreak of leprosy at Herekino amongst the natives, several of whom died from it. Others are in a fearful state. Their fingers and toes rot off, their nose, teeth, and jaws are corroding, and their bodies are rotten. He has reported the matter to the

Government. Strange to say, it got this length before we heard of it. The natives are scared, and avoid each other. Timoti Puhipi recommends the lepers being deported to the Three Kings' Island.'"

A petition by Dr. Bakewell of Auckland, New Zealand, formerly superintendent of the Leper Asylum, Trinidad, for an inquiry into leprosy in New Zealand, has been presented to the Public Petitions Committee, and referred to the Government for consideration.

Dr. George Thin states that a school in New South Wales has been closed by order of the Minister of Public Instruction, in consequence of reports that some children in it are developing symptoms of leprosy. —*Leprosy, p. 247*.

OCEANA.

The *Sydney Morning Herald*, of March 27th, 1891, says :—" The spread of leprosy among the Pacific islanders seems to be going on steadily, judging from the following report, which we take from the *Samoa Times* of January 31st—' We hear (says that journal) that leprosy has established itself at Penrhyn Island, and that there are no less than ten fully-developed cases there. The doctor of H.M.S. Cordelia, which has lately been cruising in that quarter of the Pacific, confirms the statement, and is of opinion that the disease has been brought to Penrhyn from the Hawaiian Islands by a number of refugees from the latter place. The same authority also states there is a case of undoubted leprosy at Manaheke Island. These facts strengthen the argument we have used in our columns, as to the urgent desirability of steps being taken to prevent leprosy becoming an established institution in our midst.'"

NEW CALEDONIA.

In the French penal colony of New Caledonia, leprosy has made its way with fearful rapidity. Previous to 1853 leprosy was unknown in the colony, but recently two leper asylums have been established. In a report on leprosy, presented to the Parliament of Victoria by the Secretary to the Board of Health, Sydney, it is stated that the Board of Health has received a communication from His Excellency the Governor of New Caledonia, M. Pardon, to the effect that the disease is extremely prevalent in that Colony, where about 500 of the native population are affected, and seven persons of European parentage, six convicts, and one child, all French, have been officially reported as suffering from the malady.

The *Journal of the American Medical Association*, March 22nd, 1890, says:—"Leprosy is reported to have found its way to New Caledonia, the French penal colony, and already there are hundreds of cases among the natives and convicts."

The *British Medical Journal* of April 25th, 1891, has the following:—" Dr. M. A. Legrand has recently published an account of the introduction of leprosy into the French convict settlement of New Caledonia which is at once interesting and instructive. In 1846, when missionaries first landed in the island, the disease was entirely unknown there, nor did it exist in 1853, when the French formally annexed it. In October, 1880, M. Vauvray, chief of the Health Department, sent back five negroes from the New Hebrides on the ground that they had lost their fingers and toes from leprosy. He at the same time requested the authorities to prevent the introduction of lepers into the colony. In September, 1883, M. Brassac reported that there were several cases of leprosy in the north part of the island, and suggested the establishment of a lazaretto for their reception. The authorities, however, took no steps, and in 1888, Dr. Forné, chief physician and president of the Committee of Hygiene, presented an elaborate report, in which he stated that the cases of leprosy could then be numbered by hundreds, especially in the north. But the committee adopted the ostrich-like policy familiar to such bodies, and it was not till the following year that, yielding to the force of public opinion, the executive decided to establish two leper houses, one at Pic des Morts, near the Bay of Canala, and the other in the Isle of Goats in the Noumea roads. Forty lepers were confined in the former, and twenty in the latter. In May, 1890, the total number segregated was seventy, and there had already been fifteen deaths. A third lazaretto is about to be established not far from Houaïlou. Though it was not until 1883 that the first cases of true leprosy among the aborigines of New Caledonia were observed, more or less legendary accounts of earlier appearances of the disease are current among them. Thus a Chinaman, covered with hideous sores, is said to have arrived in 1866 or 1867, and to have lived for several years with a native tribe, several of whom afterwards developed disease of the same nature. Whatever may be the true history of the importation of leprosy, there can be little doubt that it was imported, and at the present time, according to M. Legrand, it exists everywhere in New Caledonia, and has acquired a foothold in the great majority of the native tribes. Europeans have also suffered. The course of the disease appears to be more rapid than elsewhere, a fact which M. Legrand attributes to the habit which the natives have of scarifying the maculæ and the tubercules, often to a considerable depth, with pieces of glass, and to their ruthless use of caustics. M. Legrand considers that these barbarous therapeutics, together with tattooing and burning with moxas, which seems to be their fashion of expressing affliction at the loss of relatives, have much to do with the spread of the disease. He explains the ravages made by the disease in virgin soil like New Caledonia by the fact that the people, not being aware of the danger, take no precautions against it."—*Archives de Médecine Navale, February, 1891.*

In all the French colonies vaccination has been prosecuted with rigour, and has been followed by the increase of leprosy, just as in England the increase of infantile syphilis is due to arm-to-arm vaccination, as shown in the third report of the Royal Commission on Vaccination. The barbarous therapeutics, the tattooing and burning, have existed among the natives from time immemorial. Vaccination has been but recently introduced.

FRANCE.

M. Besnier, a member of the French Academy of Medicine, has reported that leprosy, far from disappearing by degrees, is spreading rapidly. Since the extension of the French colonial possessions, soldiers, sailors, traders, and missionaries, have fallen victims to it in large numbers.—*British Medical Journal*, *October 22nd, 1887, p. 919.*

According to the London *Evening Standard*, October 26th, 1891, Dr. Besnier reports the number of lepers in Paris at 100, there being at the St. Louis Hospital eight persons afflicted with this disease.

SPAIN.

"Leprosy has been on the increase in different parts of Spain for some years past, and the extension of the disease has at last aroused the attention of the Government. On February 16th the Director-General of Beneficence and Sanitation sent a circular letter to all governors of provinces calling on them to take such steps as may seem necessary under the circumstances."—*British Medical Journal, March 5th, 1892.*

In a communication to the *Lancet*, January 16th, 1892, "On the Origin and Spread of Leprosy in Parcent, Spain," founded upon investigations by Drs. Codina and Zuriaga, Dr. George Thin introduces the following table and comments :—

LEPROSY NOT DUE TO HEREDITY.

"*Table Showing the Cases of Leprosy in the Towns referred to in this Report.*

Towns.	No. of inhabitants.	Date of Invasion.	Cases up to 1887	Lepers Existing. Males.	Lepers Existing. Females.	Lepers Existing. Total.	Remarks.
Parcent	150	1850	65	21	7	28	I gathered the data stated in the present table during my visit to the towns in the district of Parcent. Although I have endeavoured to obtain my information as accurately as possible, I am unable to guarantee its correctness. The towns sometimes hide the truth as to the number of lepers existing, and I may perhaps have made some mistakes; but if there are any, they will consist in showing too small, rather than too great, a number of lepers.
Laguart	400	1868	31	20	5	25	
Pego	1200	17—	—	14	6	20	
Orba	160	1873	25	12	6	18	
Pedreguer	720	1809	79	9	3	12	
Murla	120	1870	14	8	2	10	
Sagra	100	1848	13	5	3	8	
Benidoleig	80	1869	7	4	1	5	
Gata	440	1860	8	3	1	4	
Jalon	560	1867	8	2	1	3	
Denia	650	17—	—	3	1	4	
Oudara	325	1862	7	2	1	3	
Tounos	80	1860	10	2	1	3	
Beniarbeig	130	1871	6	3	—	3	
Sanet	90	1884	3	1	1	2	
Alcahali	98	1870	6	2	—	2	
Jaica	320	—	—	2	1	3	
Benichembra	120	1872	3	1	—	1	
—	—	—	285	114	40	154	—

"There is nothing in the soil, occupation, food, or race to account for any difference in the number of lepers which are to be found in these towns respectively. It also shows that the proportion of lepers to the population of the towns is not connected with the length of time that the disease has lasted, and therefore is not in relation to the opportunities given by heredity, even if it were assumed that heredity was a cause. Parcent, which is the most striking example, shows in twenty-seven years, in a population of 150 inhabitants, 65 cases of leprosy, of whom 28 were living at the end of that period; whilst Pego, with 1200 inhabitants, and where the disease has lasted since last century, had only 20 living lepers. Pedreguer, in which we know there was leprosy in 1809, with a population of 720, had in about forty years 79 lepers, of whom 12 were living at the date of the report; whilst Murla, with only 120 inhabitants, had had 14 cases in seventeen years, of whom 10 were living at the date of the report.

"Excluding heredity as an insufficient cause of these cases, and as otherwise being discredited, the difference of the rate of increase of leprosy in these similarly situated villages is best explained by the assumption that the opportunities for contagion have been greater in some cases than in others, even if we did not have the statements which I have collected from two independent sources— namely, from the Mayor of Parcent, referred to by Dr. Zuriaga, and from Dr. Codina's report to the Director-General at Madrid. Another sad fact comes out from a study of this table—namely, that in many of the towns the appearance of the disease is comparatively recent and that in this part of Spain leprosy is spreading. The necessity for a hospital in Parcent seems to have been realised at last, for we find that a commission visited the neighbourhood in June, 1887, for the purpose of finding a site, and were offered one by the municipal corporation free of cost."

No inquiry appears to have been made, either by Dr. Codina, Dr. Zuriaga, or Dr. Thin, as to vaccination being a possible cause, which, according to a communication to me from Senor U. Montez, the Spanish Consul in London, has been obligatory for many years. This gentleman writes (London, May 26, 1892):—"Apart from previous ordinances on the subject, the law making vaccination obligatory on the whole of Spain is dated the 28th of November, 1855." This mode of propagation, where the contaminating virus enters directly into the blood, is surely more credible than the one suggested by Dr. Thin, of contagion (simple contact), unless Dr. Thin, like other pathologists, interprets the word to include inoculation and vaccination.[1]

THE UNITED KINGDOM.

The following appears in the *British Medical Journal:*—"Dr. T. Colcott Fox, one of the Honorary Secretaries of the Dermatological Society, has been good enough, in reply to an inquiry, to forward for publication the following list of cases of leprosy shown to that Society since its foundation:—

DATE.	EXHIBITOR.	WHERE FROM.
July 12th, 1882,	Mr. Hutchinson,	India.
,, ,,	Dr. E. B. Baxter,	Dutch and Chinese parentage.
Oct. 11th, ,,	Dr. Crocker,	Singapore (shown at International Medical Congress).
Dec. 13th, ,,	Dr. Liveing,	India.
Mar. 14th, 1883,	Mr. Hutchinson.	?
,, ,,	,,	?
Oct. 10th, ,,	Mr. M. Baker,	Antigua.
,, ,,	Dr. Sangster,	? If English case. Sections of nerves.

[1] Baron, in his "Life of Jenner," vol. i., p. 604, says that Mr. Allen, Secretary to Lord Holland, writing to Jenner from Madrid in 1803, observes:—"There is no country likely to receive more benefits from your labours than Spain; for, on the one hand, the mortality among children from small-pox has always been very great; and, on the other hand, the inoculation for the cow-pox has been received with the same enthusiasm here as in the rest of Europe." . . . The result, however, was the reverse of satisfactory; the writer adding, that "the inoculation of the spurious sort has proved fatal to many children at Seville, who have fallen victims to the small-pox after they had been pronounced secure from that disease."

Jan. 9th, 1884,	... Mr. Baker,	... ?
Feb. 13th, ,,	... Dr. Crocker,	... ?
,, ,,	... Dr. Stowers,	... India.
Mar. 13th, ,,	... Dr. Allchin,	... ?
July 11th, ,,	... Dr. Crocker,	... ?
Jan. 9th, 1885. ...	,,	... India.
May ,,	... Mr. Hutchinson,	... ?
July ,,	... ,,	... Cape Colony.
July 1887,	... Mr. M. Morris,	... ?
,, ,,	... Dr. Crocker,	... ?
Oct. 10th, 1888,	... Dr. Cavafy,	... ?
Feb. 13th, 1889,	... Dr. Crocker,	... ?

"Dr. Fox adds that he has also seen the following private cases since 1879:—

1. Boy from Demerara.
2. Girl from Demerara.
3. Man from Cape Colony.
4. Man from India.
5. Lady from Orange Free State.
6. Lady from Honolulu.
7. Lady from Cape Colony.
8. Officer from Bengal.
9. Man from India.

"Dr. Fox adds that he has seen three cases in hospital practice, but all these have also been under the care of other physicians at other hospitals.

"Dr. Larder has now two cases in the Whitechapel Infirmary."—*British Medical Journal, March 30th, 1889.*

SOUTH AFRICA.

In consequence of the serious increase of leprosy following the imposition of vaccination in Cape Colony, a Select Committee was appointed by the Legislature in 1883 to take evidence as to the cause of such increase, but, strange to say, no inquiry was made, nor were any interrogatories submitted to the witnesses, as to vaccination being a possible factor in the case. Mr. T. Louw, M.P., was appointed Chairman of the Committee.

Dr. Henry Anderson Ebden, President of the Medical Board, testified that he had resided many years in India—the Punjaub, the Himalayas, Rajputana, Western India, where leprosy is prevalent, and in other parts. He has been in the Cape, consecutively, close upon 22 years, and during that time had seen a great deal of leprosy. He was sure it was on the increase.

> Q.—"As you are inclined to think that the disease is contagious, do you think it dangerous to the health of other people to use food prepared by leper hands?"

"I should be sorry to see a leper cook, and I go further than that. In vaccinating, I think hardly a medical man would take vaccine lymph from the arm of a leper infant. I know it has been our practice for the last twenty years not to do so."

The Rev. Canon James Baker testified that "the increased spreading of the disease in many parts of the colony is now generally admitted. It is spreading among both the white and the coloured races, especially in places near the sea coast."

In the appendix E to this report, is a letter from Dr. Wm. R. Turner, dated Vrendenburgh, 1st Sept., 1883, in which I find the following:—"Leprosy in parts of Saldanha Bay is spreading so rapidly that, if some measures are not at once taken by the Government, all the surrounding districts will probably become infected. I know of more than twenty cases in one place alone, in every stage of the disease, and am sorry to say it is not confined to the coloured portion of the inhabitants."

The conclusion stated in the official report is "that leprosy prevails extensively in this Colony, and is steadily spreading among both white and coloured classes."

In June 11th, 1884, an Act was passed by the Legislative Assembly, Cape of Good Hope, entitled, " The Leprosy Repression Act, 1884," with the following preamble :—" Whereas the disease of leprosy is prevalent in this Colony, and has lately been spreading, and continues to spread, and it is desirable to check the extension of such disease, and, if possible, to exterminate it."

At a public meeting in Cape Town in December, 1890, Sir Gordon Sprigg said that there were between 600 and 700 lepers in Cape Colony.

In a paper read before the Epidemiological Society of London by Dr. Phineas S. Abraham, p. 3, I find the following :—" In the South African Reports for 1886, a decided increase of leprosy is stated to have taken place in the districts of Alexandria, Bedford, Clanwilliam, Herschel, Malmesbury, Paarl and Stockenstran ; and for 1887 the spread of the disease is reported not only from most of these districts, but also from Wynberg, Stutterheim, and Kokstadt. The majority of the medical officers speak very strongly on the subject. For example, one of them (from Alexandria) writes that 'leprosy is certainly spreading rapidly, and unless some active and efficient measures are soon taken, it will become a matter for the most serious consideration.' Another (from Bedford)—' I believe it to be considerably on the increase, and should be stringently dealt with.' Another (from Malmesbury)—' With reference to leprosy I cannot but repeat my statements of the last years to the effect that the disease is slowly but surely increasing, each fresh case acting as the nucleus to a more or less extended infection.' Another (from Paarl)—' It is deplorable to see what strides it is making.' And in 1887, ' As for leprosy, although it is making rapid strides, there is no notice taken of it' (*i.e.*, by the authorities). More than one of these district surgeons, indeed, assert or imply that the Boards of Management, and others who have the power to put the Act in force, shirk the duty on the score of expense, and that the Public Health Act is to some extent a dead letter."

In consequence of the continued increase of this disease, another select Committee of Inquiry was appointed in 1889, under the presidency of Chief Justice Sir J. H. de Villiers. In the appendix J to the Report, printed by order of the Legislative Council, pp. xiv. and xv., the Rev. Canon J. Baker, F.L.S., F.S.Sc., says:—" I entertain no doubt that leprosy is spreading in this colony at the present time. Observant and intelligent persons have assured me that they have recently met, in various parts of the colony with more cases than in previous years.

"There are many and great difficulties in getting correct information on the subject. I have known patients to be carefully concealed, and the relations of the affected do not like to be spoken to as to the mode of contraction of the disease. I have given great offence by calling attention to particular cases. I believe the number to be much greater than is known by medical practitioners, or by the Government authorities."

From the minutes of evidence it appears that Dr. H. C. Wright, district surgeon at Wynberg, being under examination, testifies as follows :—

Q. 5. "Will you state roughly the number of cases in your district?"—" About twenty; but it is impossible to state exactly. A great number of cases are concealed. I have not the slightest doubt that there are more cases than have come under my notice. There are, for instance, some cases I suspect to exist, because I am aware that leprosy is in the family, and lately some of these people have disappeared ; they are never seen, and, I believe, are hidden away."

Dr. Simons, District Surgeon of Malmesbury, was asked :—

Q. 75. "Since your appointment as District Surgeon, have you known any increase in the number of persons affected?"—" Yes ; certainly."

Q. 76. "Is the disease principally confined to coloured persons?"—" No ; it is not confined to coloured persons. I know of several cases where families of white farmers are affected."

Dr. W. H. Ross, Police Surgeon in Cape Town, was asked:—

Q. 230. "During the time you were Police Surgeon in Cape Town, did you meet with many cases of leprosy?"—"Yes; in going about, I used to see about a dozen a day among the poorer classes."

The Hon. Dr. Atherstone, M.S.C., who has practised in the colony 50 years, chiefly in Graham's Town, where he was District Surgeon for 26 years, and who has always taken a great interest in the subject, said:—

Q. 341. "I am decidedly of opinion that it (leprosy) is spreading."—July 18, 1889.

The following are *Extracts from the Report of the Select Committee on the Spread of Leprosy*, President, Sir J. H. de Villiers, July, 1889, p. 8:—

"The result of the inquiry has been, in the first place, to establish the fact that leprosy is on the increase in the colony. Many of the District Surgeons say that, in their particular districts, there is no such increase, and others again are unable to express any opinion upon the question, but in the more populous districts of the colony, such as the Cape and the Paarl, and even in some of the outlying and less populous districts, such as Alexandria and Stockenstrom, the District Surgeons report a marked increase in the number of cases. It should be borne in mind that the victims of this loathsome disease naturally endeavour to conceal it from others, as much and as long as possible, and that many more cases are sure to exist than have come under the notice of the medical men, whose answers have been received, or whose evidence has been taken. Your committee estimate the number of lepers in the colony to be upwards of 600."

On page 12 is an extract as follows :—" The committee conclude—(1.) 'That leprosy is on the increase in this colony.' (2.) 'That the disease will continue to increase unless effectual measures are adopted to check it, and if possible to stamp it out.'"

On the 8th April, 1890, the Governor of Cape Colony and High Commissioner Sir Henry Lock laid the foundation-stone of a new leper hospital at Robben Island, which will provide 200 beds for the unfortunates stricken with a fatal and repulsive disease. A writer in the *Cape Argus*, May 20th, 1891, who signs himself "Epaminondas," and seems, by the character of his frequent communications, to be much concerned regarding the inadequacy of the measures undertaken to check the ravages of the scourge, concludes his letter thus— " In the cause of humanity; for the suppression of the deadly disease; for the safety of the general bulk of the inhabitants, and, last but not least, the alleviation of the afflicted, it seems to me that it is the bounden duty of the Government to meet the question face to face, and devise some means to cope with this terrible and, if unchecked, disastrous evil now pervading the Colonies, and Transkei particularly."

In the valuable Consular Reports issued by the Government of the United States I find one for June, 1887, from Consul Siler, of Cape Town, which is not without interest and instruction. Mr. Siler says, "Not until 1845 was any attempt made by government to check or to stamp out the disease. In that year a leper asylum was established at Robben Island, seven miles from Cape Town, and, up to 1884, 744 lepers had been admitted to the institution, and comprised but a very small proportion of the leper population, as the segregation of lepers was not made compulsory. In fact, lepers mingle freely with the other citizens, and their appearance is so common that they attract little attention in the streets. At the Cape Town Fish Market I have seen lepers at work cleaning and curing fish, and the disgusting sight did not seem to deter buyers."

The rapid increase of the disease, particularly among the European population, as described in the recent reports of district surgeons, has aroused the Colonial Government to action, and a second and larger asylum is in process of construction, the present accommodation being wholly inadequate to provide for all the afflicted applying for admission.

A correspondent writes to the *Cape Argus*, May 21, 1891, and, referring to the alarming increase of leprosy in South Africa, says :—" It is now two years since the out-cry commenced, and yet, what has the Colony done for the benefit of the wretched people cursed with this insidious disease, and for the protection of their neighbours? Here, in the Transkei, nothing! It is notorious that in every magistracy in the Transkei leprosy is rife and spreading rapidly, and, sad to say, over the lepers themselves there is absolutely no restraint. They frequent the public offices and trading stations unchecked.

By their horrible hospitality, they provide lavish feasts of Kafir beer, invitations to attend the same being scattered broadcast. Fancy drinking Kafir beer, prepared and filtered by leprous hands? And this is done every month in the year!"

Under the head of "The Public Health," the South African Directory for 1891, p. 446, observes that "with a view of checking the spreading of venereal disease and leprosy, which have for some years past been reported to be phenomenally prevalent in various districts of the Colony, Acts were passed in 1884 and 1885 giving the Government powers whereby it was hoped that these disorders would eventually be stamped out." One of these is entitled, "The Leprosy Repression Act," which gives the Colonial Secretary power to forcibly remove and incarcerate any known leper in a leper asylum or hospital. This drastic measure had not then been promulgated or put in force. There are many lepers in good positions in the Colony, of both white and mixed races, whose friends would make any sacrifice rather than have them segregated with their fellow-sufferers, either in Robben Island or elsewhere. Those who advocate the forcible deportation of lepers from their friends (most of whom would willingly keep them from public observation in thinly-settled districts) can know little of the heart-breaking scenes constantly witnessed at the separation of these afflicted persons at Honolulu when about to undergo perpetual banishment to Molokai, Hawaii. Although a terribly repulsive and loathsome disease, leprosy is not communicable by simple contact. The Colonial legislature would have served the cause of the public health more effectually by directing their attention to municipal sanitation, and discouraging the practice of vaccination, which, according to the opinion of district surgeons, and the best informed authorities in South Africa, has been instrumental in largely spreading both syphilis and leprosy. It is hardly possible for a disinterested observer and inquirer to come to any other conclusion.

The Cape Town *Times* of March 5th, 1892, says: "A correspondent writes to the Orange Free State *Express*, under date February 25 : The first batch of these unfortunates (the Theba 'Nchu contingent), twenty in number, were despatched from here on Saturday last, and probably twelve or fifteen more will leave here within a few days. The lepers are all natives, mostly Baralongs. It was hard work for the Landdrost, with the aid of the mounted police, to hunt them up. It would take too much time to relate all the cunning (in this case excusable) devices resorted to by the families of the lepers to evade the law. It was a heartrending sight to see how mothers and children parted. It must be done, for the sake of the general safety, but it was an awful spectacle to see and hear the cries of distress, especially of the sound relations who remained behind."

Referring later to the same deportation of lepers, the Port Elizabeth *Telegraph*, March 12th, says—"Although the large number of fifty-four lepers have been despatched to Robben Island, it is believed that others have managed to conceal their condition from the authorities." . . . "It is a very remarkable fact that, whilst in the last census the number of lepers were returned (from this district) as four, already fifty-four have been discovered suffering from the loathsome disease of leprosy."

Dr. Alexander Abercromby, author of "Thesis on Tubercular Leprosy," writing to me from Cape Town, April 20th, 1892, says the disease is now "spreading rapidly amongst the white population and better class of people."

INDIA.

In the speech delivered at Marlborough House, London, June 17th, 1889, the Prince of Wales stated that one of the chief centres of Leprosy is India, where there are 250,000 lepers, and that our colonies contained unnumbered victims to this loathsome disease. The *British Medical Journal*, September 13th, 1890 (p. 639), reports that "a comparison of statistics regarding lepers during the thirty years 1851-81, shows that their number has been increasing in India at the rate of about 30,000 every ten years. During the last ten years the rate of increase is supposed to have been higher." I have before me communications from staff surgeons, medical officers of health, superintendents of leper hospitals, and medical practitioners, showing the spread of leprosy in various provinces of India, and in other countries. The Rev. G. M'Callum Bullock, of the London Mission, Almora, writing 21st August, 1889, says:—"It is the general opinion of residents, both European and native, that leprosy has increased in Kumaon during the past thirty years, and there are upwards of 1600 lepers in Kumaon alone out of a population of $1\frac{1}{8}$ millions."

Dr. C. T. Peters, in his report on cases of leprosy treated at Belgaum, Presidency of Bombay, dated

Bombay, June, 1879, says:—"Judging from Mr. M'Corkhill's figures, there were not less than 22·8 per cent. of the population, in the Belgaum districts alone, afflicted with some form or other of leprosy."

In a paper read before the Calcutta Microscopical Society, December, 1890, Dr. W. J. Simpson said it was certain that leprosy was on the increase, an opinion confirmed in a letter to me, dated August, 21st, 1889, by Dr. Chunder Ghose, Medical Superintendent of the Leper Asylum, Calcutta.

The City Coroner of Bombay says that leprosy is vastly increasing in that city. The *Times of India*, February 21st, 1891, estimates the number of lepers at large in Bombay, at 1000. At a meeting of the Municipal Council of Bombay, reported in the *Times of India*, April 12th, 1889, various speakers describe the terrible state of things existing in the city. Mr. Kirkham saw lepers near the public tank dressing their terrible sores, scratching their ulcers against the iron railing of the Elphinstone High School, where the boys sat on coming out of school. Dr. Blaney said, "all over Bombay, in dark corners, in gullies where rats and bandicoots had taken their abode, these lepers were hiding themselves, thrown out by their families, to pine away neglected and forlorn."

At a meeting of the General Committee of the "Homeless Leper Relief Fund," Bombay, held at the Municipal Rooms, the President, Sir Dinshaw M. Petit, said that the hospital (which contains over 200 patients) was overcrowded, and further admissions had to be refused. Having no homes or places of refuge, lepers hang about the bazaars of the large cities in India, forced by their necessities to sell fruit or vegetables, and to expose their maimed bodies to the gaze of the public, in order to obtain a wretched living.

Referring to the newly opened Matoonga Asylum, Bombay, Mr. Commissioner Acworth writes, May 26th, 1891 :—" With accommodation for 190, I had yesterday 226 inmates, but fortunately a new ward has just been completed, and this overcrowding will temporarily cease, though only temporarily. If I had room for 500 I could fill the asylum in a week."

The *Times of India*, May 21st, 1892, says :—" While the Matoonga Asylum is seriously overcrowded with lepers, and there are, besides, between forty and fifty bad cases in the Byculla Leper Dhuramsala, Sir Dinshaw's lakh of rupees and the land for the extension of the asylum lie still idle because of the deadlock between the Government and the Municipality over the police charges question. As the Government decline to budge in this matter, the Corporation, not altogether unjustifiably perhaps, refuses to undertake the responsibility of the Leper Asylum. Unless something is done to remedy this state of things, our streets will again be overrun with homeless lepers, and Mr. Acworth's labours in the cause of these afflicted people will practically be brought to naught."

The *Lahore Civil and Military Gazette*, May 30th, 1891, in a graphic narrative of the suffering caused by leprosy, bearing the signature " A. H.," in the leading article column, observes :—"A great deal has been said and written on the subject of the lepers and leprosy by people who have seen and pitied the miserable condition of native lepers, who parade their affliction before the public in our streets and thoroughfares, soliciting alms from the passer-by. The majority of English ladies and gentlemen who are told such persons are lepers understand and know so little about the horrible disease that they are inclined to regard them as ordinary crippled beggars, afflicted with a disease peculiar to natives, and from which Europeans are happily exempt. This is far from being the case: leprosy seems to have obtained a terrible hold over our white brethren and sisters in India, many of whom are hiding away, alone and forgotten, in the thickly populated slums and by-lanes of our large cities. I could conduct my readers to godowns and huts where English men and women are to be found in Calcutta in a horrible condition ; some in the last stage of the disease."

In a Presidential address on the "Geographical Distribution of Diseases in Southern India," delivered at the annual meeting of the South Indian and Madras branches of the British Medical Association, Surgeon-General George Bidie, M.B., C.I.E., speaking on the subject of leprosy, said:—" According to census returns the proportion of lepers amongst the population of Madras is 4·4 per 10,000, against 5·2 in Bengal, and 8·5 in Bombay ; but there is reason to believe that these figures fall short of the actual extent of the disease. In Madras it is on the whole slightly more prevalent in coast districts than in inland, the ratios being 4·9 in the former, and 4·4 in the latter, per 10,000 of population. The proportion of lepers in the several districts ranges from 2·0 in Coimbatore to 10·5 in Madras city. The districts showing the highest ratios next to Madras are Nilgiris 8·0, Tanjore 7·0, and Chingleput, Malabar, and North Arcot, each 6·0 per 10,000. The disease attacks Europeans and Eurasians as well as natives, but is most common in natives. The propagation of leprosy is no doubt largely influenced by heredity, but recent observations appear to show that it is also contagious. In localities in which lepers are at large with the disease in an active state, and having open sores, there seems to be an increased tendency to fresh cases amongst the general population."—*The British Medical Journal, p. 115, July 20, 1889.*

BURMA.

In a communication from Mr. C. G. Bayne, the Officiating Secretary of the Chief Commissioner of Burma, to the Secretary of the Government of India, dated Rangoon, 6th December, 1889, and published in the *Journal of the Leprosy Investigation Committee* for February, 1891, it is said that the majority of officers questioned state distinctly that, in their opinion, leprosy is increasing in Burma. Mr. Smeaton, Commissioner of the Central Division, says:—"In the opinion of the majority of the gentlemen consulted, there are more lepers now than there were ten years ago." Mr. Norton, Deputy Commissioner of Rangoon, remarks:—"Those best qualified to form a judgment on the subject are of opinion that lepers are more numerous now than formerly." Surgeon-Major Baker and Dr. Frenchman have come to the same conclusion. Mr. Bayne observes:—"The opinion of these officers is of special value, because both of them, particularly Dr. Baker, have paid much attention to leprosy, and have much experience of it. Dr. Baker gives reasons which are based on observations of actual facts, and are not merely impressions."

CEYLON, TONQUIN.

In Ceylon, as I learned by personal inquiries made in the island in January, 1891, leprosy is extending rapidly amongst the native population. The Leper Asylum at Hendala, near Colombo, one of the oldest in India, which in 1880 contained only 100 lepers, has now 208; and Dr. Meier, the resident Superintendent, does not hesitate to say that, in his opinion, the disease is steadily increasing.

There are about 200 lepers at large in the city of Colombo, and about 1800 in the island. Dr. Kynsey, the Surgeon-General for Ceylon, reported in 1885 that leprosy had decidedly increased since 1862, as the number of patients then in the asylum was 63, but had increased in 1885 to 151. Dr. Kynsey says:—"I have no doubt that a certain reproduction of the disease is going on, whatever the factors are at work, and that the proportionate growth of leprosy in the colony is by no means diminishing."

In a communication to the Government of Hawaii, Dr. Kynsey remarks that leprosy is not confined to any community, but is more frequently observed among the Singhalese and Tamlins; seldom among the Eurasians, and more rarely among Europeans, and is chiefly found among the poor, ill-fed, ill-housed classes of the community. The Eurasians, I may observe, as well as the better-class Europeans, absolutely decline to be vaccinated from native lymph sources, to which the native population are obliged, reluctantly, to submit.

An Anti-Leprosy Association has recently been organised in Bengal by certain benevolent members of Hindu communities in the Presidency. Their efforts are being directed especially to ameliorate the condition of the lepers in the Santhal Parganas, forming the southern portion of Bhagalpur, a very poor district, where the people can do little to help either themselves or their afflicted neighbours.

The *British Medical Journal* gives an account of a leper village near Hanoi, Tonquin, where, out of a population of 400, nearly one-half are affected with leprosy. The lepers of Hanoi doubt the contagiousness of leprosy, and the chiefs of the village affirm that there has not been a single case of contagion.

CHAPTER II.

IS LEPROSY CONTAGIOUS?

ONE of the most debatable points in connection with the spread of leprosy is that of contagion, and amongst dermatologists there are rival schools—contagionists and anti-contagionists. In the report of the Committee of the Royal College of Physicians, issued in 1867, thirteen were in favour of contagion, and thirty-four physicians and experts in various parts of the world were convinced that the disease was non-contagious. The chief authorities in Norway, including Boëck and Danielsen, who had forty years experience, were opposed to contagion; and this is the prevailing view in Norway at the present time.

Dr. G. A. Hansen says:—"If people wash themselves, and take the least care of themselves, when they come in contact with lepers, I do not think there is any danger whatever. It is a remarkable fact that not one of the nurses or servants in our Asylums (Norway) has caught the disease, although they daily wash and dress the patients."

In the pursuit of my investigation, I have been confronted on every hand by the most conflicting theories with regard to the causation of leprosy, and particularly with regard to this question of contagion. The contagionists, when pressed, I found invariably included inoculation, and interpreted the word in that sense. They admitted that the leprous discharge might be touched with impunity, when the integument is intact, but not otherwise. Every nurse, doctor, attendant or laundress, in the hospital, is bound to come in repeated contact with pus from ulcerated tubercles. It is only by the insertion of the leprous virus into the blood, through a sore, prick or abraded surface, that the disease is communicable. This view is now held by the highest authorities in all parts of the world. At the same time, there are others who hold that the disease is transferable in a lesser degree by inhalation, heredity and cohabitation.

From personal inquiries made at asylums and lazarettos in various countries where leprosy is endemic, I am convinced that, apart from the risk of inoculation, there is little or no danger of contagion, using the word to mean simple contact between unbroken surfaces of the body. So far as my investigations have extended, the only country where the belief in communication by simple contact prevails to a certain extent is Hawaii; but here also I found much diversity of opinion, not a few using the word contagion to include inoculation, both accidental, as in a cut or a sore, and by design, as in vaccination.

I believe that the instances of communication apart from inoculation of this disease (if they exist at all) are extremely rare, but the theory is opposed to the results of most inquiries.

A medical resident of sixteen years' standing in British Guiana told me that the disease was being extensively disseminated in some unexplained and mysterious way, as the infected population had greatly augmented of late years; you encountered them in churches, at balls and public meetings, in the streets, and the market-place. Several leprous patients were pointed out to me at the Colonial Hospital, in close proximity to the other inmates, and I may observe that only the worst cases (and these belonging to poor families) are segregated at the Leper Hospitals. The lazarettos at Gorchum and Mahaica, British Guiana, at Trinidad and Barbadoes, were full to overflowing; new wings were in progress, or had recently been added, and the demand considerably exceeded the present accommodation in each establishment. No one appeared to be afraid of contagion, and I could not learn of a single case so communicated. After going through the various buildings of the Leper Asylum at Mucurapo, Trinidad, and seeing the unfortunate patients in every form of this hideous and mutilative disease, I said to the lady superintendent (of Dominican Sisters), who had been in charge of the institution for seventeen years, "Have you no fear of contagion?" "Not the slightest," she promptly replied. "And you and your assistants do all that conscientious nursing requires?" (This includes washing the sores and bandaging the limbs of the unfortunate inmates.) "Certainly, and feel it a joy and privilege to be of service to these afflicted people." "Has any case of infection by contact to doctor, nurse, attendant, or laundress ever been reported during your superintendence?" "Not one."

This experience was confirmed at the lazarettos in Barbados and elsewhere; and some of the nurses and attendants have been employed from ten to thirty-two years.

At the leper asylum in Ceylon, I learned that the laundry work had been managed by one family for three generations, and no case of infection had ever been recorded of laundress, nurse, or doctor. Similar experiences were related to me in South America, South Africa, and at the leper asylums in Norway.

The officials connected with the leper settlements at Molokai, and the Hospital of Suspects, Kalihi, near Honolulu, where I saw some of the worst cases, have not the slightest fear of contagion. They told me that they had never known a medical attendant or nurse contract the disease by simple contact. Of inoculation through sores, or wounds in the skin, or the entrance into the blood through the minutest prick or abrasion, a wholesome fear is entertained, not only amongst the native population, but by the officials; and not without sufficient reason, as will be seen by the facts detailed in another chapter.

The British Consul in Crete, in a memorandum to Baron Ferdinand de Rothschild, M.P., on the subject of leprosy in that island, concludes that the disease is not contagious, from the fact that there are "several cases of healthy women married to, and living with, lepers for years without being in the least affected. In fact, if the disease were decidedly contagious and hereditary, it would inevitably spread much more than it does, considering that the lepers are perfectly at liberty to marry among themselves or with healthy persons, and that their children remain with them like those of other people, without any precaution being taken on their behalf."

Dr. Arthur Mouritz, in his official report to the

Honolulu Board of Health, dated Molokai, February, 1886, says:—"The washerwoman for the hospital at Kalowao (Molokai) has washed the soiled clothes of the worst cases, certainly many of them so, in the settlement for the past seventeen years."

In a communication by Dr. Van Deventer, Director of the Suburban Hospital, Amsterdam, to the Hawaiian Government, the writer says:—"Not one case of contagion has ever been recorded."

Dr. Trousseau, of Honolulu, who told me he devoted much attention to the causation of leprosy in Hawaii, says:—"Is leprosy infectious or ever contagious in the proper sense of the word, that is, by contact mere and simple? I emphatically say 'No.' I am supported in that opinion by the whole medical world, and by my personal experience."

Dr. Manget, formerly superintendent of the Leper Asylum, British Guiana, observes:—"My own opinion is in favour of the contagiousness of leprosy, and that it may be propagated by the matter of ulcerated tubercles being applied to any raw surface; but I admit that I have met with cases which would seem to preclude the idea that the disease can be considered contagious in the ordinary sense of the term."

In the Leprosy Committee Report of 1887, signed by Dr. C. Handfield-Jones, chairman, it is stated:— "The committee believe that leprosy is not contagious in the conventional sense of the term, but, if at all, is only so in low degree and under exceptional circumstances."

Dr. Max Sandreczi, director of the Hospital for Children, Jerusalem, says:—"I am obliged to declare that the result of my researches gives me the conviction that leprosy is by no means contagious, and that consequently the exclusion and isolation of the patients is both a useless and a cruel measure."—*Lancet, Aug. 31, 1889, p. 423.*

The *Lancet*, June 22, 1889, p. 1252, says:—" There is hardly an hospital in London that has not had within its walls cases of leprosy within the past decade—inpatients, it is true, who have contracted the disease in countries where it is indigenous. Nor, so far as we know, has there ever been an instance of the communication of the disease from one of those subjects to others in this country."

Mr. Jonathan Hutchinson, F.R.S., LL.D., in answer to the questions published in No. 1 *Journal of the Leprosy Investigation Committee*, " Is leprosy contagious?" suggests to inquirers into this subject the following important considerations :—

" A certain number of lepers arrive every year in England from abroad. They usually remain in England and are allowed to mix freely with their friends. Children are permitted to go to schools, married couples continue co-habitation, inmates of hospitals and workhouses are, unless specially loathsome, placed in the general wards ; in brief, not a single precaution against contagion is ever taken, and yet the disease never spreads. Precisely the same statements are true of French practice. It is believed that there are sixty lepers in Paris at the present time, and I am told that two leading Paris surgeons have each a leper employed as a household servant. Yet the disease never spreads in Paris any more than in London.

"The officers of leper hospitals, surgeons, nurses, and students, hardly ever become the subjects of the disease. . . . In Norway I believe that no instance of an official becoming a leper has ever been known, although the exposure has been most free."

And in a footnote to his article, page 74, Mr. Hutchinson adds :—" Surgeon-Major Porteous, in 1855, published a list of servants who had been employed in the Leper Hospital, Madras. It included eleven servants who had been employed in the wards in periods varying from ten to fourteen years. None of them had become the subjects of leprosy."

Dr. Van Someron, who had charge of the hospital six years later, says :—" There is no record of any of the medical officers connected with the lazarettos having become affected with the disorder, nor have I heard of its ever having attacked the attendants of those who in private families were its victims."

Dr. W. Munro, the author of a work on leprosy, explains his views of contagion as follows :—" I do not pretend to express any distinct belief as to the probability of the disease being conveyed by simple contact, being more inclined to believe that it is carried by inoculation in most cases."

The *Lancet* of June 28, 1890, referring to the theory of contagion in connection with the spread of leprosy, says:—" But there are conditions and limits to the contagion : probably it occurs only through inoculation." This opinion is supported, according to the *British Medical Journal*, October 11, 1890, in the despatch from the Government of India relating to the isolation of lepers. It particularly notes that many of the medical authorities in India consider that the evidence at present available goes to show that leprosy is contagious only in the sense that it is inoculable. The "Report of the Royal College of Physicians on Leprosy," issued in 1867, states "that the all but unanimous conviction of the most experienced observers in different parts of the world is quite opposed to the belief that leprosy is communicable by proximity or contact."

Sir Erasmus Wilson says leprosy is endemic, but not contagious.

Dr. William B. Atkinson, secretary of the State Board of Health, Philadelphia, in diagnosing the case of John Anderson, a Swedish leper, observes that there is no danger of contagion except by inoculation.

Dr. Shoemaker, of Philadelphia, says leprosy is only contagious through inoculation.

In a study on leprosy, based on personal observation, Dr. L. Duncan Bulkley has arrived at the conclusion that the disease is not in any proper sense of the word contagious, but there is reason to believe that under certain conditions it can be inoculated.—*Family Doctor*, *June 11, 1892.*

Dr. H. M'Hatton, Macon, Georgia, concludes a paper " On the Propagation of Leprosy," published in the " Transactions of the Medical Association of Georgia," by stating that it is non-contagious, and quotes the report of the committee of the English College of Physicians to the effect that, out of sixty-six answers to their inquiries, only nine speak of it as contagious, forty-five as non-contagious, and twelve are silent.

Dr. John L. Mears, Medical Superintendent of San Francisco Board of Health, says—" Although this disease (leprosy) may not be contagious in the ordinary acceptation of that term, we are satisfied that it is communicated by inoculation."

Drs. Fox and Graham report in the transactions of the American Dermatological Association for 1883, page 197, as a result of their combined investigations, that leprosy is contagious by inoculation, and *there is no reason for believing that it is transmitted in any other way.*

Dr. P. W. Farrar, Nevada, Iowa, W.S., in a communication to Mr. L. F. Andrews, Secretary, State Board of Health, February 21st, 1885, p. 205, says:—" Leprosy is not contagious in the usual acceptance of the term. It requires actual inoculation of pus or blood into the circulation through open vessels or abraded surfaces, and there must be favourable cachectic condition to the action of the virus."

Dr. Bevan Rake in a communication to the Acting Surgeon - General, dated Maraval, 11th July, 1889, says:—" In a paper received from St. Louis, Missouri, only the other day, I saw that Dr. Bockmann estimates that there must be in Minnesota about 100,000 persons of Norwegian descent whose ancestors were lepers; and yet leprosy never appears amongst them; all the leprous Norwegians in the State are imported, so that leprosy does not appear to have spread there either by heredity or contagion."

In an article entitled " Notes on Leprosy as observed in Antigua, West Indies," Mr. John Freeland, Government Medical Officer, observes:—" On the subject of contagion I certainly agree with Mr. Hutchinson when he says that the profession divides itself into two camps, one asserting contagion and the other denying it; but I think that the contagionists, or those who believe that contagion takes an important share in the spread of the disease, are, in this part of the world at least, in a decided minority. No one, I imagine, would absolutely deny that contagion might be artificially effected by inoculation; but the chances of such an event happening accidentally are so remote that it can hardly be taken into account."—*British Medical Journal, October 5th, 1889.*

Dr. James H. Dunn, Professor of Dermatology in the Minnesota Hospital College, in a clinical lecture on leprosy, reported in the *North - Western Lancet*, March 1st, 1888, said:—" The question, Is leprosy contagious? has been a source of much discussion and contention. At times and in some countries it has been looked upon as markedly contagious. Some writers still regard it so; but at the present day the great majority of dermatologists teach that it is not, at least not in the ordinary sense of the term. There is no evidence to show that the malady has in any instance spread by contagion in a country where leprosy is not endemic."

Mr. T. H. Wheeler, the British Consul of Bogotá, South America, in his Report for 1890 to the Foreign Office, No. 804, observes that although public opinion favours the belief prevalent in Colombia that leprosy is contagious, in the climates of Tocaima and Agua de Dios it is not so:—" For more than one hundred years that these places have been the chosen resort of lepers in all stages of the disease, who have mixed freely with the other inhabitants of the district, there is no case on record of the disease having been contracted by contagion."

Dr. Alfred Ginders, in a communication to the Inspector-General of Hospitals, etc., Wellington, New Zealand, on Leprosy among the Maoris, dated Rotorua, 4th July, 1890, states his opinion that the disease is not infectious or contagious in the ordinary sense, but "that, in all probability, the worst cases have arisen from direct infection of the blood by inoculation, either accidental or premeditated." The only premeditated form of inoculation in vogue is that induced by the lancet of the vaccinator.

The Medical Superintendent, Leper Asylum, Calcutta, Dr. Madhub Chunder Ghose, in his Report to the Honourable H. Beverly, President of the District Charitable Society of Calcutta, 27th August, 1889, says :—" It seems to me, after an experience of fifteen years in the asylum, that leprosy is not contagious or infectious in the proper acceptation of the term. Recently I have taken the full history of all the lepers in this asylum, and, with one or two exceptions, the origin of the disease could not be traced to contagion ; some acquired the disease from an hereditary taint, some from the effects of syphilis, and the indiscriminate use of mercury ; but in most of the cases the origin of the disease could not be satisfactorily traced, but I have no doubt that the disease can be communicated by an abraded surface absorbing leprous matter."

Dr. Ghose adds :—" To prove my assertion as to the non-contagiousness of leprosy, I beg to bring forward the following facts, that is to say, my own personal experience of the disease for over fifteen years. There is an inmate of the asylum, by the name of Doris, who is a non-leper, and who has been at the asylum for over twenty years, sleeping in the same ward, constantly mixing with the lepers, eating with them, etc., and he has not contracted the disease.

"There is also an idiot boy at the asylum, a non-leper, who has been an inmate for over ten years; he also sleeps, eats, and mixes freely with the lepers; this boy, also, has not the slightest trace of the disease.

"The dhoby attached to the asylum, with his father and grandfather before him, have washed the clothes of the lepers for more than thirty-five years; none of those showed any signs of the disease.

"The native doctor, Runchanun Dass, who lived with his family for over ten years in the premises of the asylum, neither contracted the disease, nor did any of his family.

"The dressers, Buddye and Narain, acted, the former for twelve years, and the latter for ten years : they did not suffer from the disease.

"The dressers, Rajjian and Jaddao, have been attached, off and on, the former for eight years, and the latter for ten years (this man is yet at the asylum as a dresser), and I have recent news regarding Rajjian, who has gone to his country : these men are unaffected. The dressers have, daily, to handle sores, wash unhealthy ulcers, apply ointment, etc., besides having to shave the lepers periodically. The present Christian cook and his father have been working at the asylum for over twenty years. The father died a non-leper, and the son is free of the disease.

"Other cooks, who work for a few years and then go to their country, have never been attacked. The sweepers, Roohon and Bustee, have worked more than seven years without contracting the disease. Both the men have been discharged, and are yet living. Other sweepers, who have been working a short time each, also have not suffered.

"The Durwana have not contracted the disease. The present Durwan has been now over five years in service.

"I have myself been attached to the asylum now for over fifteen years, visiting the lepers daily, cutting and handling them, without having suffered.

"My predecessor, Dr. K. Stewart, was in medical charge of the asylum for over twelve years, and remained free of the disease till his death. My assistant, Dr. H. W. Mitnish, M.R.C.S., England, has been at the asylum for over eight years, and is healthy."—*Report presented to the Hon. H. Beverley, M.A., dated Calcutta, 27th August, 1889.*

Dr. Vandyke Carter, of Bombay, says:—"I have not met with any evidences of the contagious nature of leprosy that bear sifting."

Dr. Day, of Calcutta, who, according to Dr. Balehandra Krishna, L.M. and S. of Bombay, has made leprosy his special study, says, in the *Indian Mirror*, that he does not believe in the contagious nature of leprosy.

Dr. J. Jackson, Bengal, in reply to a communication from the Royal College of Physicians, writes:—"It is not contagious in the ordinary sense of the term. . . .'
—*Leprosy Report, p. 202.*

When Mr. Commissioner Acworth, of the Bombay Municipal Corporation, paid an official visit to the Madras Leper Hospital, he was informed that the sweepers employed to wash the ulcers of the lepers did not contract the disease, although some of them had been doing the work for fourteen years.

Mr. A. Mitra, L. R. C. P., L. R. C. S. (Edinburgh), Chief Medical Officer, Kishmir, in an article in the *American Journal of Medical Sciences*, Philadelphia, 1891, observes:—" Of course, contagion by inoculation is possible, and often takes place in various ways. In India, people usually have their feet and skin bare, and, therefore, there is every likelihood of inoculation."

Mr. A. Mackenzie, Secretary to the Government of India, writing from Simla to the Minister of Foreign Affairs, Honolulu (October, 1885) says:—"On the whole, it is believed that the medical evidence tends to show that the disease is not contagious. In support of this view, it may be mentioned that not a single servant of the asylum at Almora, in the Kumaun District of the North-Eastern Provinces, appears to have contracted the disease during the thirty-one years for which there is information."

Dr. W. A. Kynsey, the Surgeon-General, Ceylon, says: —" It (leprosy) is not considered contagious in Ceylon, and lepers are not generally shunned by their relatives or friends for fear of infection, but are often maintained by them in their own houses. It is, in my opinion, not contagious as syphilis, Parangi, the exanthematous diseases. There is no conclusive evidence in the hospital records of communicability by direct contact with, in close proximity to, diseased persons. The attendants of the hospital have for years been in close association with lepers in all stages of the disease, the head-servant for more than twenty years; and the washing of the establishment has been performed by a family in the neighbourhood for four generations; but not a trace of the disease, as I have reason to know, has been observed among them."—*Leprosy in Foreign Countries, p. 9.*

Dr. Dixon, Medical Superintendent of Robben Island, Cape of Good Hope, in a report published in the *Journal of the Leprosy Investigation Committee*, No. 3, July, 1891, says:—"The evidence gathered from officials and patients long resident on Robben Island shows that there is no authentic instance, with possibly one exception, of any non-leprous person on the island having contracted the disease, either direct or indirect, with the leper residents."

Dealing first with the possible exception, the circumstances were as follows:—A lad, son of the shoemaker, constantly associated with the lepers; he ate of food given to him by them, and was in the habit of fishing with their tackle; it is stated that on one occasion, when using the lepers' tackle, *he had a wound on his finger.* For about ten years he exhibited symptoms, said to be those of leprosy. He died in 1888, having suffered for about ten years. It cannot be held that there is conclusive proof that this solitary case originated by contagion.

The evidence of the older officers goes to show that, until about the year 1884, all the lepers' soiled and filthy linen was washed by the female lunatics in cold water only, and was often mixed with the underclothing of the lunatic patients in the process of washing. This practice was probably in vogue for upwards of thirty years, yet there is no alleged or recorded instance of any lunatic patient contracting leprosy on Robben Island.

Dr. W. H. Ross, Police Surgeon in Cape Town, was asked—

Q. 252. "You are aware that the bacilli have been in the saliva of lepers, would not that render the disease liable to be spread by the act of kissing?"—"Not unless there was some cracked surface on the lips or mouth. I have never known of a case of leprosy having been contracted on the island, although they mix there freely." — *Report of Select Committee on the Spread of Leprosy, Cape of Good Hope, July, 1889. Minutes of Evidence.*

Mr. Davidson of Madagascar says:—"Leprosy is contagious by inoculation only."

Dr. W. V. M. Koch, the Acting Superintendent of the Leper Asylum, Trinidad, writing on the subject of contagion, explains that "the entrance of the (leper) germ into the system will take place if it is brought into contact with an absorbing surface—any abrasion of skin or mucous membrane being sufficient for this purpose."— *Surgeon General's Report for 1891, p. 71.*

Dr. Alexander Abercromby writes to me from Cape Town, April 20th, 1892, that after thirty years experience he holds that leprosy is partly contagious, and explains that he does not use the word contagion in the strict sense, "but when there is a discharge from a leprous sore, and this coming in contact with the tissues of a healthy person will develop the disease ; or the saliva of a person coming into contact with a slight abrasion of cuticle, or healthy mucous surface."

Under the head of " Leprosy in Havana," the *British Medical Journal*, June 18th, 1892, says that the number of cases in the Real Casa Hospital de San Lazaro at the present time averages 80 to 90, but seldom reaches 100. In 32 cases (40 per cent.) no family history of the disease could be obtained. "In no single case could leprosy be traced to contagion, and of the 25 persons employed in the hospital, only one (a chaplain) contracted the disease during the last 12 years. Dr. Arango, the present medical superintendent, has never known any case in which the disease could be distinctly traced to contagion, and he knows persons who have lived twenty-eight or thirty years in the hospital without contracting it."

CHAPTER III.

LEPROSY COMMUNICABLE BY INOCULATION.

WHILE the preponderance of medical and scientific opinion is against the theory that leprosy is, in the ordinary sense of the word, a contagious disease, the evidence in favour of its being communicable by inoculation is overwhelming. Even those who strongly uphold the theory of contagion invariably include inoculation as one of the principal means of communication. However widely authorities differ as to the other causes to which leprosy is attributed, such as climatic influences, unwholesome and putrid food, want of salt, a fish diet, malaria, heredity, contagion, syphilis and insanitation, we may safely affirm that there is a practical consensus of opinion as to its inoculability.

To the question, Is leprosy inoculable? Sir William Moore, K.C.I.E., late Surgeon-General, Bombay Staff, Hon. Physician to the Queen, says :—" Professors Damisch and Kobner proved by an experiment that leprosy may be communicated to animals by inoculation. There is also the well-authenticated case of a boy, Miller, who pricked himself with a needle used by a leper, from which injury leprosy developed. Then there was a case of a medical student pricking himself when performing a *post-mortem* examination on a leper."

Sir William Moore has himself cited (*Journal of the Leprosy Investigation Committee*, No. 1, p. 28) a case of inoculation in a person with an injured hand who was employed to rub sulphur ointment on leprous patients, his family being quite free from the disease, and no history of previous association with lepers being obtainable. He considers that the sulphur ointment had no protective influence. "All that is required is the transmission of leprous discharge, which contains the microbe or germ of leprosy, to the healthy body. But in order that the poison may act it is necessary that it should come in contact with an abrasion or sore on a healthy skin. An infinitesimal portion of leprous discharge is quite sufficient."—*Leprosy and Leper Houses, pp. 2 and 3.*

In a communication to the secretary of the Leprosy Fund, dated June 2, 1890, Sir William says:—" Leprosy has been attributed to the following causes :—(1) Climatic heat ; (2) unsanitary conditions ; (3) want of salt; (4) vegetable diet; (5) fish diet; (6) lime in water; (7) malaria. But facts show that none of these are the causes." After giving the reasons for this conclusion, the writer adds :—" My views that leprosy is a phase of inherited syphilis, communicable, however, by inoculation, and the reasons for such views, have been expressed in my 'Manual of the Diseases of India,' 1887."

This view accords with the most eminent medical authorities in all countries where leprosy prevails. In a report on Leprosy in Cyprus, by Dr. Heidenstam, chief medical officer for the island, transmitted by the High Commissioner, Sir Henry Bulwer, to Lord Knutsford, and presented to both Houses of Parliament, March, 1890, the author, after dismissing various theories put forward to explain the spread of leprosy, such as heredity, the use of putrid food, salt pork, mal-hygiene, malaria, and miasma, says :—" My researches have led me to the conclusion that leprosy is what should be termed an inoculable disease, inasmuch as the virus is transmitted into the system in like manner as many other maladies, notably syphilis, anthrax, glanders, etc.; but it has not the same action in all constitutions, nor in all circumstances of life, and is of a long and slow incubation. My further researches and studies have not in any way altered the opinion I then expressed, and I am more than ever convinced that the direct cause of leprosy is simply and solely due to the inoculation of the virus of a person affected into another up to that time free.

"It has been advocated that instances of the communicability of leprosy have been rare, and so doubtful that it is impossible to rely on their authenticity. In this island, at least, I have met many cases where the slightest doubt could not be entertained."

Dr. Olavide, of Madrid, at the Paris Dermatological Congress of 1889, maintained that leprosy was evidently a parasitic disease, contagious, and inoculable. This authority refers to a curious fact, that it is observed by preference amongst the soldiers and monks who have resided previously in America and in the Philippine Islands. In Spain, while vaccination is not always carried out amongst the civil population, it is rigorously enforced in the army, as I was informed some years ago by the then British Consul, Mr. M'Pherson, of Madrid.

At the same Congress, Dr. Zambaco-Pach, communicating the results of his inquiries concerning leprosy in the Isle of Mitylene, mentions that while he is an anti-contagionist, the discovery of the leprous bacillus has somewhat shaken his ideas. He records the curious fact that, of about 120,000 souls in Mitylene, 15,000 are Mussulmans, and amongst these there is not a single leper to his knowledge. Dr. Zambaco omits to note the fact that Mussulmans in most countries have a rooted aversion to, and distrust in, vaccination, and escape the ordeal whenever they can.[1]

Dr. Sutherland, of Patna, observes that:—" Another test of the prevalence of leprosy in this district was to ascertain the proportion of leprous persons in Patna gaol. Among Mussulmans, two were affected, or one in twenty-eight; while among the Hindoos there were seventeen persons affected, or one in every sixteen."

" This is certainly a startling assertion, one person in every sixteen affected with the taint of leprosy, and yet it comes from the pen of a careful observer."—*Leprosy a Communicable Disease, by Surgeon C. N. Macnamara, p. 15 and 16.*[2]

[1] This repugnance was made known to me by means of personal investigations in Ceylon in 1890-91. "The First Triennial Report of the Working of the Vaccination Department in Bengal" mentions, amongst other races, the Mahomedan Ferazis, who display the utmost repugnance (as do also the higher classes in India) to vaccination.

[2] The *Madras Times*, Nov. 28, 1891, referring to the opposition to vaccination, which, the editor observes, is by no means confined to the ignorant populations, says:—" Official reports show notably that the Lubbays and Mahomedans, as a class, resist vaccination, and do much to prevent the authorities from tracing the age of children over six months, with a view to enforcing the Act. Male vaccinators, it is remarked, are prohibited from entering their houses on the ground that these are *gosha*, and Hindu female vaccinators are unable to cope with the difficulty satisfactorily. It has been proposed to employ a Mahomedan female vaccinator, but a suitable woman who can read, write, and serve notices has not yet been found."

In the *Archives de Médecine Navale et Coloniale*, September, 1890, is an article by Dr. F. Forné on the "Contagiousness of Leprosy," of which theory the writer is a firm adherent. Dr. Forné says:—"One of the arguments invoked for the purpose of denying the contagious character of leprosy consists in saying that the persons attending on the lepers—religious persons, infirmary attendants, medical men—do not contract leprosy. It is important to refute that assertion by showing that it is contradicted by facts." The writer then proceeds to give details of the case of Father Damien, who, after sixteen years' residence at the leper settlement at Molokai, succumbed to this disease, as Dr. Forné supposes, through contagion. Of this there is not the slightest proof, unless contagion is understood to include inoculation, and that this is intended would appear by the following reference to another case to illustrate the author's theory:—

"Dr. Hulin de Goden, medical officer of the leper settlement of Desirade (French Antilles), says that Sister A. became leprous after having pricked herself in the fingers of her hand with a sewing-needle while mending articles that had been used by lepers. Whether it was after having pricked her fingers matters little, since it is by the skin of the hands that the leprous contamination would have taken place in both cases. He observes that, in general, the washing of linen is habitually practised under the supervision of the Sisters, while sewing, on the contrary, is more often carried out by themselves.

"The accidental inoculation would have taken place before 1878, the date of the first leprous manifestations;

in 1881 the affection had taken so considerable a development that Dr. Hulin de Goden decided to isolate the patient.

"An analogous fact has been observed in Tahiti, where leprosy has been transferred by passing from the native race to the white race, absolutely in the same manner as we have the fact produced in the Sandwich Islands, in New Caledonia, and in Guiana.

"A religious woman from Europe, free from hereditary taint, who was working at the linen-drapery of the hospital at Tahiti, inoculated into herself the terrible malady with a sewing-needle under the same conditions as the religious woman of Guiana. She was sent back to France in 1885 as being affected with leprosy."

Dr. Woods, cited by Dr. Hahn, expresses himself thus :—" In the hospitals of Calcutta, and other districts of the East Indies, hospital attendants, positively free from all hereditary taint, have contracted leprosy by means of accidental inoculations undergone in the exercise of their functions ; Dr. Robertson, Director of the Leper Settlement of Seychelles, became leprous during his period of service in that house."

Dr. Forné says : " Dr. Hillebrand cites the following case :—At Borneo a young European boy was in the habit of playing with a child of colour affected with leprosy. One day this latter plunged the point of his knife into an anæsthetised part of his body, an operation which was immediately repeated by his comrade with the same knife. Some time afterwards the European started for Holland, there attained his maturity, and, at the end of nineteen years, returned to Borneo fully affected with leprosy."

In a memoir read by Besnier before the Academy of Medicine, Paris, October 11th, 1887, Leloir, a high authority, is stated to have pointed out that lepra would appear to be contracted by inoculation and not by contact, from the fact that the epidermis does not, as a rule, contain bacilli, and that the epidermo-dermic basal membrane apparently constitutes a barrier to the passage of the micro-organism in either direction.

One of the latest contributions to this important subject is an able work entitled "Leprosy," by Dr. George Thin (London, 1891), in which the question of contagion is fully discussed. The author has succeeded in bringing together (pp. 139-166) the largest number of examples of alleged contagion of any writer I have met with. Some of these are distinctly traceable to inoculation by means of sores, pricks, gunshot wounds, abraded surfaces. The following are amongst the examples:—

No. 14.—Dr. Duncan, Civil Surgeon, Julpaiguri district, states that a healthy woman sustained a gunshot wound in the thigh. She had no leprous relations, but her husband was a leper with ulcerations on his hands. The woman became a leper.

No. 53.—A young coloured boy, while suffering from an eruption, played with a boy who was a leper; the previously healthy boy became a leper about a year afterwards, whilst his family remained untainted.

No. 67.—Dr. W. H. Ross cites a case which occurred in a European family without any leprous taint. J. K., while playing with leper boys, pricked himself with one of their fishing hooks, and became an undoubted leper.

On page 162 Dr. Thin remarks: " No one doubts that syphilis is a contagious disease, because surgeons, nurses, and attendants may fulfil their duties for many years in Lock hospitals without becoming infected; and it is a matter of every-day experience that a member of a large family may pass through all the infectious stages of syphilis, living in constant association with brothers and sisters, without the disease being transmitted ; yet it is quite certain that in all those cases the disease could be communicated by an inoculation of the simplest kind." From this illustration we infer that Dr. Thin considers leprosy transmissible by inoculation, and sometimes uses the word contagion in that sense, as I have noticed with many other writers on leprosy.

On page 66 Dr. Thin quotes Dr. Donelan as authority for a case of leprosy due to inoculation.

Professor Cayley, F. R. C. S., says *(Journal of the Leprosy Investigation Committee*, p. 36) that leprosy is directly inoculable.

Dr. John Murray, Inspector-General of Hospitals London, in a communication to Dr. P. S. Abraham, Secretary of the National Leprosy Fund, June 9th, 1890, says: "I consider that it (leprosy) is communicable from the sick to those that are well, probably through a broken surface, as an ulcer or wound, and that it may be communicated by inoculation."

Dr. Liveing, physician to the Middlesex Hospital, in his Gulstonian Lectures for 1873, says:—" Facts, too, are slowly accumulating which tend to prove that the casual inoculation of leprous matter is one actual means of spreading this fell complaint." Dr. Hoegh, in his "Report on Leprosy" for 1855, quoted by Liveing, "suggests that the disease is communicable through the *Itch Acarus*,[1] which in Norway commonly infests the skin of lepers."

In his handbook on the "Diagnoses of Skin Diseases," 1880, p. 284-5, Dr. Liveing writes :—" Leprosy has within the last thirty years been imported and spread rapidly amongst natives of certain islands where it was before quite unknown. It is probable that in a certain stage of the disease it is inoculable. This appears to me the most reasonable explanation of its progress amongst a new population."

In reply to a communication which I addressed to the superintendent of the Leper Asylum, Bergen (which institution I visited in 1889), Dr. G. Armauer Hansen, the Physician General of Leprosy in Norway, the discoverer of the *bacillus lepræ*, says :—" I think leprosy to be inoculable ; I, moreover, think that leprosy in most cases is transferred by inoculation."

Sir Erasmus Wilson, F.R.S., in his work on "Cutaneous Disease," 1864, says :—" Lepra is a blood disease. The origin of the disease is doubtless an animal poison, but the source and nature of the poison are unknown. One remarkable case lately under our observation has led to the belief that it may be communicated by inoculation."

INDIA.

Dr. Balchandra Krishna, L.M. and S., in his pamphlet entitled " Leprosy in Bombay in its Medical and State Aspects," suggests a mode of reconciling the conflicting opinions as to the contagious nature of leprosy. He quotes the *Lancet* of June 29, 1889, " The discovery of

---
[1] Presumably the *acarus scabiei*.

the *bacillus lepræ* by Hansen has greatly strengthened the belief in contagion. This bacillus has never been found in any disease or condition other than leprosy, while it has invariably been found by competent observers in the skin of tuberculated lepers from all parts of the world. In the non-tuberculated cases, on the other hand, it is invariably absent in all the sores in the diseased nerves. But it has been found in the nerves themselves when the disease is not of too old a standing, and in those skin lesions of mixed nerve and skin leprosy not dependent upon disease of the nerve trunks. This goes far to explain the non-transmissibility of nerve leprosy. The evidence, on the whole, then, is strongly in favour of the disease being communicable; but all are agreed that it is only so in the *ulcerating stage* of the *tubercular* or the *mixed form* by inoculation of the pus on an abraded surface, either directly from the patient, or from the stained clothing or other objects contaminated by the secretions from the leprous sores. It is certainly not communicable by aerial infection in the same way as small-pox or other exanthemata."

"This seems to me," says Dr. Krishna, "to explain satisfactorily the reasons why some observers found cases which convinced them of the contagious nature of the disease, while others met with some which did not show any contagious nature. It also explains the three cases which I have mentioned above, as on no other hypothesis can they be explained."

In a paper on leprosy in the *Times of India* (Bombay), August 13, 1889, Dr. Balchandra Krishna says that the evidence in proof of the communicability of the disease from man to man is overwhelming.

OTHER MEDICAL AUTHORITIES. 107

Brigade-Surgeon H. V. Carter, of Bombay, referring to the spread of leprosy, says: "The direct communicability of leprosy is at least a good working hypothesis."

Surgeon-Major Pinkerton, in his evidence before the Royal Commission on Vaccination, testified that leprosy was increasing in the cities of India, and believed that it was inoculable.—*Second R.C. Report, p. 6*.

In a letter to the *Times*, June 12, 1889, Surgeon-Major Pringle, late of the Sanitary Department, Her Majesty's Army, Bengal, refers to the danger of spreading leprosy by both inoculation and vaccination. "The fact is, the amount of the virus of leprosy with which Father Damien was unknowingly fatally inoculated might have been, and probably was, very minute. I am amply justified, from a careful study of smallpox inoculation and vaccination during the whole of my thirty years' Indian service, in stating that, unless prompt and stringent measures are taken in Bombay, leprous inoculation will become far more possible, and hence probable, than it may appear at present."

Dr. Joq. Frank Periera, Medical Superintendent of the Leper Asylum, Bombay, India, in a communication to the *Times of India*, November 18, 1890, gives his opinion that the contracting of leprosy is mainly due to its inoculation by means of open sores from one person to another, and adds: "In most, if not in nearly all the cases treated by me, their previous histories have, almost without exception, disclosed the fact of the disease being due chiefly to heredity and inoculation."

Dr. Cunningham, the special adviser to the Government of India, admits the principle of the inoculability of the disease.

As to the supposed cause of leprosy, "my experience," says Dr. S. M. Shircore, of Moorshedabad, India, "does not tend to the belief that this disease is contagious in its nature, unless by direct inoculation."

Dr. H. A. Ackworth, Municipal Commissioner, Bombay, who has devoted much attention to the leprosy question, writes to me from Bombay, 24th May, 1891:— "All the medical men that I have met have agreed that in whatever of any other methods leprosy may be transmissible, it is certainly so by inoculation."

Dr. W. K. Hatch, M.B., Surgeon, Bombay Army, reports in the *British Medical Journal*, June 26th, 1886, p. 1713, that on June 27th, 1885, a student, while making a *post-mortem* examination on the body of a confirmed leper, cut his left forefinger at the top, and received a small abrasion, which resulted in certain characteristics of leprosy. This writer has been promised further particulars.

Dr. Neve, of Kashmir, says that in leprosy the *bacillus lepra* is always present; that the period of incubation is so long, "that a few positive instances of inoculation or contagion outweigh an immense amount of negative evidence."—*Leprosy, by Dr. George Thin, p. 62.*

Dr. G. D. M'Reddie, Civil Surgeon, in his letter to Dr. Ghose, dated Hurdor, the 18th February, 1888, states:—"From observations I know leprosy is hereditary. It is also contagious in the sense that it is necessary for the discharge from a leprous ulcer to come into direct contact with the broken skin of the recipient, or the blood of a leper to be inoculated into the system, as in vaccination."—*Report on Leprosy to the Hon. H. Beverley, M.A., by Madhub Chunder Ghose, Leper Asylum, Calcutta, August 27th, 1889.*

Surgeon C. N. Macnamara sums up the question of the communicability of leprosy as follows :—" The arguments, therefore, against the communicability of leprosy do not refute those in favour of it ; consequently, I can arrive at no other conclusion than that leprosy is communicable; but it is necessary for the propagation of the disease by this means that the discharge from a leprous sore should enter the tissues of a healthy person, and, further, the disease may even then (unless under peculiar circumstances) remain undeveloped in the system for years."—*Leprosy a Communicable Disease, p. 43.*

SOUTH AFRICA.

In Appendix A to the "Cape of Good Hope Report of the Select Committee on the Spread of Leprosy," 1883, is an interesting communication from the Rev. Canon James Baker, dated Kalk Bay Rectory, August 10, 1883, as follows :—" My own opportunities for investigation have been rather exceptional, and my advantages considerable. In early life I was a student of Medicine, and subsequently of Chemistry and Natural Philosophy, at University College, London. My appointment as chaplain to the Lunatic Asylum and General Infirmary on Robben Island, where I remained nine years, put me in the way of getting experience among lepers, and I commenced at once and continued to make the nature of this terrible disease a special subject of inquiry. In my present sphere of duty I see, unhappily, many cases for investigation. The increased spreading of the disease in many parts of the colony is now generally admitted ; it is spreading among both the white and the coloured races, especially in places near the sea coast.

"Leprosy is not to be compared with small-pox or scarlet fever, as to contagion, any more than typhoid fever, but this can be conveyed in excreta, finding its *nidus*, or seat of incubation, in one part of the body only; and so, while many will escape leprosy who handle patients as they may handle decomposing corpses, with the skin of their hands unbroken, others may have their blood-vessels or absorbents come in fatal contact with the active poison of the disease."

In the body of the same report I find the evidence of a number of witnesses of wide experience, who have given the subject much attention. Hon. Dr. Atherstone, Member of the Legislative Council, and F.R.C.S , Eng., who has practised in the Colony fifty years, chiefly in Graham's Town, where he was District Surgeon for twenty-six years ; also Consulting Physician of the Albany General Hospital, and President of the Leprosy Inquiry Committee of 1889, testified as follows :—" I have formed a very decided opinion as to the nature of the disease, and the manner in which it is transmitted from one to another, and spread all over the country. Recent microscopic investigation has established the fact that the diseased tissues and secretions are invaded by numerous parasitical, rod-like organisms called bacilli, always of the same form and size, no matter from what part of the world the leper comes, or what part of the body is examined, whether the tubercles, lymphatic glands, cartilages, or suppurating sores. This specific bacillus of leprosy is no doubt the true cause, and it is spread by inoculation, either by direct contact with the secretion, or suppurating sores of the leper, or transmitted by the clothes, utensils, pipes, etc., containing these parasitical germs of the disease."

Q. 345. "You are then of opinion that it is contagious?"—"Yes; in the mode I have described; not in the ordinary sense of the word."

On page 8 I find the following:—"Another result of our inquiry has been to remove any doubt that might previously have existed as to the contagiousness of the disease. Your committee are satisfied that where the disease has not been derived by heredity from one of the parents or grandparents, it has in every instance been contracted by means of contagion. It is quite possible that the disease may not be communicable except to a person having some wound or abrasion in the skin; but when it is borne in mind that the victims often suffer from a discharge of matter from the hands or other limbs, it is not difficult to conceive how readily the disease may be communicated to persons coming in contact directly or indirectly with the sufferer."

Dr. Abercrombie, member of the Cape Town Medical Board, says (Answer 6a):—"It would be communicated to a person who came in contact with a leprous person if he had a sore or an abrasion. For instance, if he were to touch a leprous person with a sore finger, use the same knife and fork, or drink out of the same glass."

Sir Samuel Needham, superintendent of the Old Somerset Hospital, says:—"I do not think it is contagious, except in cases of cohabitation, when persons are reduced to a low state of health, through being badly fed, or when they get a cut or wound inoculated by contact with a leprous patient."

Dr. W. H. Ross, twenty-two years Police Surgeon at Cape Town, in reply to Question 311, "Do you know the case of a little boy who contracted leprosy on the island?" said:—"I know the case. The boy had no leprous relations. He associated with lepers, and one day, being out fishing with them, he ran a fish hook into his finger. Leprosy shortly afterwards made its appearance."

In the "Report of the Select Committee of the Legislative Assembly on Leprosy," July, 1889, are the following answers from the same witness :—

Q. 346. "I know of several cases in which the disease was communicated by inoculation."

Q. 351. (1) "It has been conclusively shown that it is always accompanied, if not caused, by a specific bacillus or bacilli, distinctly recognisable under high powers of the microscope in all the stages, in the skin, tissues, glands, secretions, nerves, and bones of the parts affected, undergoing ulceration and destructive degeneration by the pressure of interstitial tubercular deposits. (2) These specific rod-like bacilli, like other fungoid growths of the lowest type of generative life, such as dry rot in old wood, etc., it appears, can only gain access to the system in persons in a low state of vitality, either from poverty and filth, defective nutrition, or depressed nervous energy or constitutional debility, from heredity, or other causes of enfeebled condition of health, admitting of the growth and reproduction of the germs of these low organisms. (3) In such weakened state of the constitution, hereditary or acquired, these bacilli may be introduced into the system by direct contact with any abraded absorbent surface of the skin or mucous membranes, as in wounds, sores, pricks, etc., or with the mucous surfaces of the lips, mouth, nose, eyes, etc., but chiefly in parts distant from the centre of circulation, as in the toes, fingers, etc., where they may find a lodgment from the diminished vitality of the part being insufficient to destroy them, although sufficient to prevent its rapid growth."

Dr. J. C. Taché, Titular Professor, Laval University, Visiting Physician of the Tracadie Lazaretto, New Brunswick, Canada, reports to the Hawaiian Government ("Leprosy in Foreign Countries," 1886, p. 142-3):—
"There is a case, the facts of which are established beyond the possibility of cavil, in which the disease appears to me, as well as it did to those who witnessed it, to have been produced by the absorption of liquid matter discharged from the body of a woman who had been in a cachectic state from leprosy. At the funeral of that woman, the body was carried on the shoulders of four strong young men. The day was hot, and, on a sudden, liquid matter began to ooze out through a joint of the coffin, wetting the shoulder of one of the carriers. The wet, combined with the heat and the pressure of the sharp edge of the coffin, produced an abrasion of the skin of the young man. The contact of the liquid with the abraded surface lasted a part of the time of the

procession, and the whole length of the service, as it was only on his way home that the young man washed his sore shoulder, and changed his clothes. Some months after, that man, whose health had always been robust, began to feel unwell. In a short time the symptoms of leprosy made their appearance, and he died of the disease eleven years after the occurrence. There had never been any case of leprosy in his family, whose ancestral genealogy is traced for several generations back. In fact, the disease was not yet known as leprosy, being of recent appearance in the locality, and among these people. He was the fourth case in that place, the other three being the woman spoken of, the husband and sister of the woman, in the ancestry of whom there had never been any trace of the disease. The fifth case in that locality was the sister of the young man." It is noticeable that the incubation of this disease was of comparatively short duration.

Dr. Manget, Surgeon-General, British Guiana, in the "Report of the Royal College of Physicians on Leprosy," p. 45, observes:—"My own opinion is in favour of the contagiousness of leprosy, and that it may be propagated by the matter of ulcerated tubercles being applied to any raw surface; but I admit that I have met with cases which would seem to preclude the idea that the disease can be considered contagious in the ordinary sense of the term."

Dr. Charles W. Allen, attending physician to the North Western Dispensary for Skin Diseases, Surgeon to the Charity Hospital, in an article in the *New York Medical Journal*, March 31st, 1888, on leprosy, concludes that the disease is transmissible by inoculation from one individual to another.

In a remarkable article on "Leprosy, its Extent and Control," by Dr. H. S. Orme, published in the 20th volume of "Transactions of the Medical Society of the State of California" (1890), page 180, Dr. Saxe is quoted as giving the case of a physician's son who acquired the disease after inserting a pin into his leg which a little Hawaiian leper had just previously thrust into an anæsthetic patch on his own leg.

Dr. S. Kneeland, of Boston, U.S., who visited Honolulu in 1872, says:—"There can be no doubt that it (leprosy) is spread by cohabitation, and inoculation of its diseased fluids, in the same way as syphilis."—*Dr. Hillis on "Leprosy in British Guiana," p. 192.*

Dr. J. C. Graham concludes an article in the *Canada Medical and Surgical Journal*, October 1883, as follows:—"In all probability the disease is communicated solely by means of inoculation; and opportunities for such inoculation are very few indeed, unless there has been a long and intimate contact with a diseased person."

Dr. Graham here omits from his purview the universal practice of vaccine inoculation, which sometimes carries with it the sources of leprosy and of other diseases.

In a leading article in the Philadelphia, U.S.A., *Medical News* on "Leprosy in its Relation to the State," the writer says:—"From time to time we hear of lepers reaching this country, either from Norway, the West Indies, or from China, and it becomes a very important question how to deal with such cases. Up to a few years ago, opinion was very strongly against the contagious nature of the disease, but since Hansen's discovery of the *bacillus lepra* facts have been accumulating to show that the virus is a fixed contagion, communicable by inoculation, like syphilis and glanders."

Dr. William B. Atkinson, Secretary to the State Board of Health, Philadelphia, in his official report for 1890 of a case of leprosy (John Anderson, a Swede), observes, " There is no danger of contagion except by inoculation. The transmission of leprosy by inoculation or contact has been a debatable point with the profession for many years; but since the disease has been better studied, and the discovery of the *bacillus lepræ*, a minute organism found in every case of true leprosy, opinion is gradually but surely coming round to the recognition that leprosy may be communicated by the unhealthy to the healthy to a much greater extent than has hitherto been considered probable."

The *American Journal of the Medical Sciences*, October, 1882, has a communication on "The Question of Contagion in Leprosy," by Dr. White, who remarks that heredity as the only, or as an important factor is out of the question. " It would have required several generations to have accomplished such results; we must look, then, to the customs of the race as exceptionally favourable to inoculation, and as the only possible explanation. It is probable that leprosy, like syphilis, may be communicated under all circumstances by which some of the fluids and other products of the infected *foci* of a diseased person come in contact with abraded or excoriated, possibly with the uninjured surface of a healthy person. It would be necessary that the diseased products should be at the surface of the skin, or mucous membrane, and this would generally be accompanied during the process of softening by which the impermeable layers were removed. Thus the nodular form in its ulcerative stage would necessarily be the most dangerous phase of disease, whereas the anæsthetic form might exist for years with little danger of communicating itself to its surroundings. In this sense we may conclude that leprosy is contagious, and in these ways, probably, the disease mostly spreads in a family, a community, a nation."

Mr. Plumacher, United States Consul at Maracaibo, Venezuela, observes in his official report for 1890 (p. 695.) with reference to leprosy:—"I confess freely that I am not a believer in the theory of contagion properly so called. It will be easily understood that should matter from a leprous ulcer come into absolute contact with the blood of a healthy person inevitable infection would result, and leprosy be engendered, in the same manner that the surgeon at times meets his death through blood poisoning contracted in the dissection of a cadaver."

The United States Consul to Cape Colony, Mr. James W. Siler, in the official report on leprosy to his Government, dated March 24th, 1887, observes:—"This specific bacillus of leprosy is no doubt the true cause, and it is spread by inoculation by direct contact with the secretion or suppurating sores of the leper."

In a paper read before the State Medical Society by Dr. R. J. Farquharson, Secretary of the Iowa State Board of Health, Des Moines, and published in the *New York Sanitarian*, July, 1884, the author says:— "Leprosy is not contagious in the ordinary acceptation of that term. It requires an absolute inoculation of pus or blood into the circulation through open vessels or abraded surfaces, and at the same time it is assumed that we must have the cachectic conditions favourable to the action of the virus."

Dr. Wood, of the United States Navy, in the Fourth Report, Navy Department, says that the dressers in the hospitals of Calcutta and other portions of the East, positively free from hereditary disease, have in many instances developed it, under circumstances connecting the inoculation with their duties.

The *New Orleans Medical and Surgical Journal*, 1880, published a communication from Dr. T. H. Bemiss, Lahaina, Hawaii, on the introduction and spread of leprosy in these islands. "Alarmed," says the writer, "by an invasion of small-pox in 1853, a general vaccination of the whole population was ordered, and physicians being at that time very few on the islands, non-professionals aided in the work. It is charged by some that, as a natural result of the labours of the heterogeneous force so appointed, not only syphilis but also leprosy was greatly increased. In my last circuit trip in my district, I found very few adults who had never been vaccinated. This involves the question of inoculability, in my opinion the main, if not the only means of propagation, other than inheritance."

The same journal, 1888, says that "leprosy may be communicated from a leprous to a non-leprous person by means of a specific virus, which acts somewhat like the specific poison of syphilis, depending upon thin or denuded surfaces for its absorption, and which remains potent, very probably for an indefinite period of time."

Dr. R. Hall Bakewell, formerly Superintendent of the Leper Asylum, Trinidad, testified before the Select Committee of the House of Commons in 1871, that the inoculation of leprosy was proved as much as any fact in medical science.

In a paper read before the Auckland (New Zealand) Institute, July 20, 1891, and printed in vol. xxiv. of the "Transactions of the New Zealand Institute," Dr. Bakewell says:—"That bacilli exist in both leprosy and tubercle is beyond all dispute; that the bacilli of these diseases may be grown and cultivated in suitable media is ascertained as a fact respecting one of them—tubercle —and, although not experimentally proved as regards the bacillus of leprosy, yet is almost beyond doubt. Artificial nutrient materials have hitherto failed, and it is not allowable to try the only natural medium—the blood and tissues of a person living under conditions likely to develop leprosy. I have no doubt, from seeing the origin of leprosy cases, and studying several hundred cases of the disease, that it is not only inoculable, but that it spreads by inoculation or absolute contiguity, and I have no hesitation, after twenty years' consideration of the subject, in affirming again the opinion given before the Committee of the House of Commons."

Dr. N. B. Emerson, President of the Honolulu Board of Health, says:—"The great problem that confronts the Board of Health is the leprosy question, and the medical profession in the Hawaiian Islands are, I believe, unanimous in the belief that leprosy is a communicable disease, and a transplantable disease, communicable by inoculation."

In a communication to the Secretary of the National Leprosy Fund, dated Dublin, 6th October, 1890, Dr. John D. Hillis, F.R.C.S.I., late Medical Superintendent, Leper Asylum, Demerara, says:—"My views on leprosy are explained in my work, 'Leprosy in British Guiana,' published by Churchill & Son. A further experience of ten years has convinced me more firmly than ever that leprosy is a communicable disease, most probably by inoculation. In tropical climates, many suffer from ulcers, excoriations, etc., which may render them more susceptible when brought so much into contact with lepers."

Dr. Arthur Mouritz, who occupied the position of physician to the Leper Settlement, Kalawao, Molokai, in 1886, says:—"The *contagium* of leprosy enters the system by inoculation at broken surfaces of the skin, fissures, or chaps, on external mucous surfaces, and possibly by punctures of insects, or the presence of parasites, scabies, etc." In his report to the President of the Board of Health, Honolulu, dated February, 1886, the doctor says:—"Some weight must be attached to the views of the foreigners themselves. They, one and all, such as are now alive, emphatically declare their belief that the disease is contagious. Some give evidence of contact (immediately followed by local symptoms—direct inoculation), infection of the whole system speedily following, this again succeeded by external manifestations of leprosy within a comparatively short period."

In his "Biennial Report to the Legislature of the Hawaiian Kingdom," session 1890, Dr. T. H. Kimball considers the fact of the inoculation or transplantation of leprosy to have been proved in those islands, and that the *bacillus lepra* carries the infection.

Dr. Ginder, who investigated cases of Maori leprosy at Taupo and Rotorua, New Zealand, in his "Report to the Inspector-General of Hospitals, etc., Wellington," dated

4th July, 1890, concludes "that in all probability the worst cases have arisen from direct infection of the blood by inoculation, either accidental or premeditated."

From inquiries made from those who were intimately acquainted with the late Father Damien, I have no doubt that in his case the disease was induced by means of inoculation of leprous virus from other patients, when he resided in Molokai, through sores on the skin. While possessed of many noble traits, this worthy and self-sacrificing missionary was conspicuous for neglect of ordinary hygienic precautions.

Dr. W. Munro, in his work on leprosy, quotes a series of cases to show that leprosy is spread by contagion, but explains in page 80 the wide interpretation he gives to this much misunderstood word. He observes "that by using the word 'contagion' I do not pretend to express any distinct belief as to the probability of the disease being conveyed by simple contact, being more inclined to believe that it is carried by inoculation in most cases, though long-continued contact even of unbroken healthy with diseased skin may be sufficient."

The doctor gives particulars (p. 84) of several cases of inoculation, all of which, he says, tend to show that inoculation is the chief, if not the only, manner by which the disease is propagated, such propagation taking place quickly only when some special circumstance, as the person being wounded, makes inoculation easy and certain. This writer does not believe there is any evidence of its being communicated in food or drinks, and the only danger from association with lepers is "*when the healthy person has any cut or sore about his hands* by which he might be inoculated."

Dr. John Freeland, Government Medical Officer, Antigua, in a communication dated Antigua, Sept. 15, 1890, says:—"When I wrote to the *British Medical Journal* in October last that no one would deny that contagion might be artificially produced by inoculation, I meant, of course, inoculation by means of the actual introduction of secretion from the leper's sore into the skin of the healthy, effected either directly by the lancet, or accidentally conveyed through the broken surfaces of the leper and the healthy coming in contact." Dr. Freeland relates how that " T. S., a healthy and robust lad, who was denied board and lodging by his relatives on account of his irregular and late hours, sought accommodation and residence in an out-room situated in a leper's yard. After a time, he received rather a severe wound on one of his feet, and I was called upon to attend him, when I naturally protested against his surroundings, and wished him to go into hospital, not only that he might have every comfort and care, but that he might also be at once removed from his diseased neighbours. He would not, however, consent to leave his house, and I continued attendance long enough to discover that my patient had, since his accident, been systematically using the same basin, the lotion, and even the very rags and bandages that were, perhaps, but a few hours removed from the ulcerated surfaces of his leprous companion. I need not tell you that the healthy wound soon developed into an intractable and sloughing sore, and was, after some time, followed by those general, but unmistakable, symptoms of leprosy, which went on progressing until the disease was fully formed, and the lad died, an ulcerated and necrosed leper."— *The Lazaretto, St. Kitts, West Indies, October 6, 1890.*

THE INOCULATION OF THE CONVICT KEANU.

Whatever doubts have heretofore existed as to the inoculability of leprosy, there can hardly be any after a dispassionate consideration of the facts connected with the experiment on the condemned convict at Honolulu. The prisoner Keanu was inoculated with leprosy by Dr. Edward Arning on the 30th September, 1884, and again in November, 1885, *after previously making a most searching inquiry as to any leprous taint in his family, and a close examination of his own body.* This examination satisfied Dr. Arning that no trace of the disease could be found in him. Every precaution was taken to secure his isolation from contaminating surroundings, and means were adopted to ensure that he was not employed outside the prison walls. On the 2nd September, 1888, Dr. N. B. Emerson, then President of the Board of Health, and Dr. T. H. Kimball, Government Physician, examined the prisoner and signed the following certificate :—

"This is to certify that we have this day carefully examined Keanu, who was inoculated in November, 1885, and we find his condition as follows :—

"Ears tubercular and considerably hypertrophied ; forehead the same ; face, nose, and chin show flattened tubercular infiltration ; mouth clean, no tubercles ; face generally presents a leonine aspect.

"Hands puffed, fingers swollen at proximal phalanges, tapering to distal phalanges ; tips of forefinger and thumb of left hand are ulcerated from handing hot tin cups of tea or coffee, indicating anæsthesia.

"Body—Back thickly mottled with flattened tubercles and the surface uneven to feel, colour of the same—a yellowish brown ; front of the body, chest and abdomen, presents plaques of tubercular infiltration of larger size than back, separated from each other by wider intervals and of a brighter colour, in some cases a ruddy pink, especially over upper part of sternum.

"Legs—The infiltration thins out as far down as the knees, there being one large bright patch on the inside of the left thigh ; legs below knees quite clean and skin smooth and even to touch.

"Feet — Œdematous ; have poor circulation ; bluish colour ; soles of feet clean.

"Seat of inoculation, outer aspect of left forearm, upper third, shows a dark purplish scar, about one-and-a-half inches long by five-eighths of an inch wide, irregular in shape, keloid in aspect, dense and inelastic.

"The tests for anæsthesia were not made. Eyes with sclerotitis, muddy and infected.

"No signs of palsy about muscles of face, orbiculares palpebrarum, hands, or forearms.

"It is our decided opinion that this man is a tubercular leper.

"N. B. EMERSON, M.D., *President of the Board of Health.*

"J. H. KIMBALL, *Government Physician, Honolulu.*"

In the spring of 1890 Dr. D. W. Montgomery, Professor of Pathology, California University, microscopically examined a piece of Keanu's skin, and discovered the *bacillus lepræ* both singly and in groups. This bacillus has been found, according to the *Lancet*, by competent observers in the tuberculated form of leprosy in all parts of the world, and has never been found in any other disease or condition. Keanu has since been sent to the lazaretto, Molokai, a confirmed and incurable leper—a punishment ten times more severe than the death penalty, and, in my judgment, utterly unjustifiable. In a letter on " Leprosy" in the *British Medical Journal*, September 24, 1887, Dr. William Jelly observes :—" I daresay the poor Kanaka convict (Keanu), had he known what leprosy is, would, without hesitation, have preferred the guillotine, the garotte, or the hangman's rope."

Regarding this official declaration, Dr. C. N. Macnamara says:—

"This report establishes unequivocally the fact that the inoculated man has become leprous; and, as he had been inoculated three years previously, there is every reason to believe that the disease is the result of the inoculation. This is very much borne out by the fact that at the seat of inoculation there is what is described as a dark, purplish scar of about one inch wide, keloid in aspect, dense and inelastic.

"The importance of a positive result like this cannot be outweighed by a considerable number of negative experiments; although, so far as we know, this is as yet the only direct experiment that has been made from a leper to a sound individual, and we do not believe that its importance is lessened by the fact that inoculations made from a diseased part of the body to an apparently unaffected part of the body of the same person have, in some cases, not led to development of the disease in the inoculated part within a comparatively short lapse of time.

"The communication of leprosy in this case confirms the views generally entertained by those who hold that the disease is contagious, the idea being that it can only be conveyed from one person to another by a direct communication of leprous tissue into the moist, living tissue of the person infected; in short, that when it is contagious, it is contagious in the same way that syphilis is understood to be contagious."—*Leprosy a Communicable Disease, p. 45.*

In an article in the April (1890) number of the *Occidental Medical Times,* Dr. Sidney Bourne Swift, Resident Physician, Leper Settlement, Molokai, reports the present condition of Keanu as follows:—"Age 70 years; weight 178 pounds; leprous infiltration beneath integument of face and forehead; tubercular enlargement of lobes of both ears, the right more than the left; ulceration of palate, and extensive ulceration of pharynx; tubercular enlargement of uvula; tubercular enlargement of alæ of nose; partial occlusion of nasal fossæ, due to leprous infiltration beneath pituitary membrane; chronic conjunctivitis and phyrigium-like growth on both eyes; almost deaf; voice hoarse, and with a nasal inflection. Anæsthesia of both hands and feet, although no pronounced enlargement of ulnar or tibial nerves; numerous tubercles distributed over the entire body, but most marked on the upper and lower extremities; three small but angry-looking ulcers on outer aspect of left leg; softened tubercle on dorsum of right foot. The hands and feet have a boggy feel, and pit slightly on pressure. His appetite is good, feels well, and looks well, and may live long enough to die of old age."

In an article on "Personal Observations of Leprosy" in the *New York Medical Journal* for July 27th, 1889, Dr. Prince A. Morrow, after describing the results of the inoculation by Dr. Arning and the development of tubercular leprosy in the convict, observes that, during his visit to Molokai, he excised a small sub-cutaneous tubercle and a portion of the underlying skin. Numerous sections of this specimen were made by Dr. Fordyce, and in all of them the presence of bacilli was exhibited under a microscope. Dr. Morrow estimates that in the Sandwich Islands "about one half the cases are tuberculous, about one-third are anæsthetic, and the remaining sixth represent the mixed form; the tubercular type is the most rapidly fatal."

Dr. F. B. Sutliff, who spent four years studying cases of leprosy in Maui, Hawaii, says, referring to Keanu's inoculation:—"This case will always stand alone; I suppose no other man will ever be purposely inoculated with leprosy. The facts in the case that point towards the inoculation as having been the direct and only cause of the disease are many and strong. Still they will be attacked by those who would rather maintain their own ideas than discover the truth."[1]

---

[1] In an article on Keanu's inoculation, in the *Occidental Medical Times*, April, 1892, Dr. Sidney Bourne Swift intimates that the case made out by Dr. Arning is inconclusive, inasmuch as other members of Keanu's family have been found to be affected with leprosy. Keanu's own son, Eokepa, aged about twenty-three years, and his first cousin, Maleka, on his mother's

A well-known medical practitioner at Honolulu gave me a photograph of Keanu, which distinctly shows the appearances peculiar to inoculated tubercular leprosy at the point of insertion in the arm, as well as in other parts of the body. And he considered the experiment an absolute demonstration of the inoculability of the disease. He also unhesitatingly expressed the opinion that the dissemination of leprosy in Hawaii was largely due to inoculation by the lancet of the public vaccinator, a most serious matter not only for Hawaii, but for all other countries where the repulsive and destructive disease is endemic. Dr. Arthur Mouritz, Medical Superintendent of the Leper Settlement, Molokai, says it is doubtful whether one per cent. of the Hawaiians would resist intentional inoculation.

LEPROSY COMMUNICATED BY INSECTS.

Both in the West Indies, and in British Guiana, I found the belief prevailing amongst the people, as well as, to a certain extent, amongst medical practitioners, that leprosy was inoculated into the blood by mosquitos.

---

side, are lepers, living in the leper settlement. Eokepa left school in 1873 on account of this affliction. These cases are by no means inconsistent with the facts contained in the reports above quoted, and it must not be forgotten that the lepra disease was first discernible at the points of inoculation. Nor can they be considered remarkable, knowing how the disease had been propagated by the vaccination lancet. In one instance reported to Queen Liliuokokalani, an entire school in Hawaii was swept away, with the exception of a single survivor, by this means. However, the case for inoculation does not rest upon Dr. Arning's experiment, but on the un impeachable evidence of numerous reputable witnesses in all parts of the world, and on the fact admitted by pathologists that, given suitable conditions, all bacterial diseases are inoculable.

Nor is there anything improbable in the idea. Sir William Moore, late Surgeon-General, Bombay Staff, "is of opinion that one of the chief sources of danger is due to flies and mosquitos. These pests of Indian life may carry enough leprosy discharge to communicate the disease to a healthy person. None of us can make sure that the fly or mosquito, which irritates by its persistent attention, has not come from a leper. A fly investigates a leprous sore or discharge, carries a particle of poison in its proboscis, or feet, and next settles on some abrasion of the skin of a healthy person!" Dr. Manson, in China, says that elephantiasis has been conveyed by a mosquito.

In Dr. Wilson's communication on leprosy in *The Lancet*, Nov. 13, 1880, p. 779, the writer says:—"Dr. Manson received some reward from the Chinese Government a few years ago, for the discovery that leprosy was caused through the introduction of a poison into the blood by the bite of the mosquito, and although little has been heard of this discovery since, the idea seems to receive support from many facts, and explains the curious occurrence of that dreadful malady in the arctic regions, where the mosquito abounds."

Dr. Albert S. Ashmead has an article on "Leprosy in Japan" in the *Journal of Cutaneous and Genito-Urinary Diseases*, vol. viii. page 220, copied into the *Journal of the Leprosy Investigation Committee* for July, 1891, in which the danger of minute inoculation by insects is referred to as follows:—"The Japanese guard carefully against mosquitos and other insects, and wherever insects most abound the most endemic leprosy is found. In addition, those parts exposed to insect foraging are the seat of primary skin lesions of leprosy, as also mucous membranes most exposed to germs in food and water."

And it may be remarked that, if leprosy may be communicated by means of mosquitos and other insects, where the inoculated virus is infinitesimal in quantity, how much greater in proportion is the danger of such contamination in vaccination? In the latter case the vaccine lymph may, and often is, taken from children where the disease lies dormant, in the incubating stage, without declaring itself by the smallest signal to the eye of the most experienced physician. In all countries where leprosy is endemic, Europeans resolutely object to be vaccinated with lymph from native sources; and, notwithstanding the law, when imported lymph cannot be obtained they and their children remain unvaccinated. As a consequence, the population of Europeans attacked with leprosy is comparatively small and, indeed, of rare occurrence, except in the case of soldiers who are subject to the military regulation of revaccination. This repugnance to native lymph on the part of Europeans in the West Indies was pointed out by Dr. R. Hall Bakewell, Vaccinator-General, Trinidad, in his remarkable evidence before the Select Parliamentary Committee of 1871, and has been referred to by Dr. Castor, of British Guiana, and other authorities.

CHAPTER IV.

VACCINATION WITH REFERENCE TO LEPROSY.

HAVING shown, on the authority of some of the most eminent dermatologists and superintendents of leper asylums, and from the testimony of those who have devoted special attention to the study of leprosy, that the disease is inoculable and spread by inoculation, we proceed to inquire whether there is evidence that this inoculation may be due in whole or in part to vaccination. When dealing with this question, I am aware that I am treading on delicate ground, inasmuch as vaccination has been lauded as an operation benign in its nature, free from peril, "the greatest discovery in the history of medicine,"[1] and, out of half-a-million prescriptions or so, the only one possessing such transcendent merits as to justify its universal compulsory enforcement. It is, moreover, considered by many ardent advocates to be impolitic to do or say anything calculated to discredit vaccination. It is hardly necessary to remind our readers that there has never been a scarcity of medical inventions which have held out similar promises. Smallpox inoculation, which, according to Dr. Moore, cost the

---

[1] Sir John Simon, late Chief Medical Officer to the Local Government Board, speaking of Jenner and vaccination, says:—" His services to mankind, in respect of the saving of life, have been such that no other man in the history of the world has ever been within measurable distance of him."
—*Temple Bar Magazine*, No. 376, March, 1892, p. 373, Article, " The Growth of Sanitary Science."

nation millions of lives, was universally accepted by the profession for the best part of a century as a discovery "highly beneficial to mankind;" and it would not have been difficult, had it been originally an English discovery, to have obtained a munificent grant from Parliament, and possibly to have persuaded the Government to make it obligatory and universal. The practice of bleeding and cupping was in vogue for at least three centuries, and the use of huge doses of mixed and noxious drugs for nearly as long a period. Jenner received a reward of £30,000 from Government on his explicit assurance that vaccination would make an end of small-pox. The evidence brought before the Royal Vaccination Commission in London, by statisticians and able pathologists, abundantly shows that, while it has had no effect in diminishing either sporadic or epidemic small-pox, it has been a prolific source of the spread of inoculable maladies, such as skin disease, pyæmia, eczema, phlegmon, and, notably, leprosy and syphilis.

THE WEST INDIES.

One of the earliest medical practitioners to call public attention to the spread of leprosy by vaccination was Dr. R. Hall Bakewell, president of the Board of Health, Trinidad, who, as Vaccinator-General of that island and Visiting Physician of the Leper Hospital, possessed unusual opportunities for observing the effects of vaccination upon the health of the people. In a communication dated 31st December, 1870, Trinidad, to the Colonial Secretary, London, Dr. Bakewell says:—"The question is not as simple as it appears. It is not a question of half-a-dozen minute punctures in an infant's arm *versus* an attack of small-pox. It is a question of performing on every child that is born into the world, and that lives to be three months old, an operation sometimes, though very rarely, fatal; sometimes, but not frequently, attended with severe illness, always accompanied by considerable constitutional disturbance in the form of fever; sometimes, in an unknown proportion of cases, introducing into the system of a healthy child constitutional syphilis, but suspected in the West Indies of introducing a poison even more dreaded than that of syphilis—leprosy. And the parent is required by law to subject his child to these evils, most of which are only possible, but one of which is certain (the fever), for the purpose of avoiding the chance of an epidemic of small-pox, which, when it does occur, may or may not attack the child.

"It may be taken as proved that the syphilitic poison may be, and has been, introduced into the system of a previously healthy child by means of vaccination. But we know that leprosy is a constitutional disease, in many respects singularly resembling constitutional syphilis; like it, attended by stainings and diseases of the skin; like it, attacking the mucous membrane of the nose, throat, and mouth; like it, producing falling off of the hair, diseases of the nails and bones; and, like it, hereditary. Why should not the blood of a leprous child, whether the leprosy be developed or not, contaminate a healthy one?

"It seems to me not merely a popular opinion, but a medical one also. In returning to Europe in the spring of this year, I met several medical men from Demerara and other tropical countries, and they all considered that leprosy might be, and is, propagated by vaccination.

"Such cases as the following are difficult to account for, unless one adopts the hypothesis that leprosy is contagious (and, if contagious, *a fortiori* inoculable). The daughter of a colonel in the army, who held a staff appointment in one of our tropical colonies, contracted leprosy while in the colony. There could be no hereditary tendency, for both the parents are English, and it could not have arisen from bad diet. The young lady herself was a creole, but I was not able to discover whether she had been vaccinated from a creole or not. When I called in to see her in England, the disease was so far advanced that nothing could be done for her.

"Four cases at this moment in this colony are whites, both of whose parents are Europeans."[1]—*Royal Gazette, Trinidad, March 1, 1871.*

Dr. Bakewell was summoned on behalf of the Government to give evidence before the Select Vaccination Parliamentary Committee in 1871, and testified as follows (Answer 3563, p. 207 Official Report):—

"There is a very strong opinion prevalent in Trinidad, and in the West Indies generally, that leprosy has been introduced into the system by vaccination; and I may say that, as Vaccinator-General of Trinidad, I found that all the medical men, when they had occasion to vaccinate either their own children or those of patients in whom they were specially interested, applied to me for English lymph; and that was so marked that in one instance a man, who had never spoken to me before, wrote me quite

---

[1] Since the above was written I have been informed by the father of one of the patients that Sir Ranald Martin, on being consulted after seeing the case, pronounced that the leprosy had been caused by vaccination.—*Foot note to Dr. Bakewell's Report.*

THE SELECT PARLIAMENTARY COMMITTEE. 135

a friendly letter, in order to get lymph from England when he had to vaccinate his own child. It is quite evident that the only reason for wanting lymph from England must be that they consider it free from contaminating the system by leprosy; because, of course, there is an equal chance, and probably a greater chance in England, of the lymph being contaminated by syphilis."

Question 3564 and Dr. Bakewell's answer (pp. 207-8) are as follows :—

Q.—" Have you had experience of any case in which Leprosy has been introduced by vaccination ? "

A.—" I have seen several cases in which it seemed to be the only explanation. I have a case, now under treatment, of the son of a gentleman from India who has contracted leprosy, both the parents being of English origin. I saw the case of a child last year, who, though a creole of the Island of Trinidad, is born of English parents, and is a leper, and there is no other cause to which it is attributable. Sir Ranald Martin, who is a great authority on these points, agreed with me that the leprosy arose from vaccination." [1]

---

[1] Since the above evidence was communicated, twenty years ago, Dr. Bakewell has availed himself of opportunities for extending his researches into the causation of leprosy in New Zealand, which have served only to confirm and strengthen his previous conclusions. In a paper read before the Auckland (New Zealand) Institute, 20th July, 1891, and printed in vol. xxiv. of the "Transactions of the New Zealand Institute," Dr. Bakewell observes that the inoculation of leprosy by means of vaccination is now exciting much attention, and he gives the results of his more recent inquiries into the subject of vaccination, citations from which will be found in the appendix to this volume.

In a State paper addressed to Earl Granville, consisting of correspondence on the subject of leprosy, presented to both Houses of Parliament by command of Her Majesty, May, 1871, Dr. Bakewell refers to a boy, aged 15, brought from Guadaloupe by Dr. Brassac at the expense of the French Government, in which the disease was attributed to vaccination from a leprous source.

No notice seems to have been taken of this startling testimony by the Select Vaccination Committee in their Report; and it was not until the publication of Professor Gairdner's remarkable cases in the *British Medical Journal*, June 11, 1887, that further inquiry was attempted. In the same year Governor Robinson, of Trinidad, issued a circular to medical men in the Colony, referring to these cases. This letter contained the query, "whether the disease (leprosy) is communicable by vaccination, lymph from healthy vesicles alone being used." It is obvious that the form in which the question was submitted was but little calculated to elicit the true facts. If "healthy vesicles" alone were used, it is clear that no disease, other than *vaccinia*, could result, and vaccination would be acquitted from a serious indictment, which was the evident intention of the experts who formulated the question. To discuss whether, and how far, impure lymph, or the arm-to-arm lymph in general use in countries like Trinidad, where leprosy is endemic, could convey leprosy, would have been to travel beyond the scope of the inquiry. To this circular twenty-seven replies were received, which may be summarised as follows :—

(*a*) Upon the matters of experience, two of the

witnesses, Dr. Alston and Dr. Chittenden, do not clearly indicate how far their experience warrants them in expressing an opinion upon the question. Of the remainder, four, Dr. de Verteuil, Dr. R. F. Black, Dr. R. H. Knaggs, and Dr. Bevan Rake, appeal confidently to their own practice. The others do not appear to have had sufficiently extensive experience to warrant a conclusion.

(*b*) As to the communicability of leprosy by vaccination, two of the writers, Dr. de Verteuil, and Dr. Bevan Rake, consider that leprosy cannot be inoculated by vaccination, if pure lymph be used. Dr. Black is of opinion that leprosy can be inoculated, though lymph from healthy vesicles alone be used. All the others express themselves with uncertainty and hesitation, or else disclaim having an opinion worth expressing upon the subject. Dr. de Montbrun, senr., says that leprosy would not be inoculable if every medical practitioner selected very healthy infants, born from healthy parents, and used only the lymph which exudes spontaneously after the puncture of the vesicles. Hardly any of the witnesses refer to the notorious difficulty, not to say impossibility, of ascertaining what families are leprous; concealment, for prudential reasons, being the rule, as admitted by intelligent inhabitants in all the leprous districts I have visited. Dr. Robert Francis Black, who has resided sixteen years at Port of Spain, Trinidad, informed me that his attention had been directed to the subject, and he had not the smallest doubt as to the invaccination of leprosy. He replied to the circular as follows :—

76 Queen Street, Trinidad,
July 16, 1887.

SIR,—I have the honour to acknowledge the receipt of your confidential circular, No. 1818, dated 12th instant, and beg to state, for the information of His Excellency the Governor, that my experience of leprosy agrees with the statements of Professor W. T. Gairdner, of Glasgow, contained in your circular, and that I am of opinion that the disease in question is communicable by vaccination, lymph from healthy vesicles alone being used.

I myself have seen two or three cases of leprosy following vaccination, and have questioned the parents closely, but failed to ascertain or detect any family taint in either. Both the parents were respectively from Africa and China, the other was of creole parentage, but all the children were born here. With reference to these facts, I may mention here that, as far as I can recollect, the periods of incubation after vaccination were from two to three years; in fact, immediately after vaccination all were seized with obstinate cutaneous eruptions. As these were casual cases, I kept no memoranda, and as they did not return I lost sight of them, they probably concluding that it was hopeless to do so. I am also of opinion, for the reasons here stated, that arm-to-arm vaccination, at least in Trinidad, where leprosy is decidedly on the increase, is bad, as many very respectable families here are tainted with the disease, and nearly all the Portuguese have some member of their families actually diseased.

In conclusion, I may mention that I am also of the opinion that leprosy, like syphilis, tubercular phthisis, and cancer, is hereditary and contagious.

I have the honour to be, Sir,

Your obedient servant,

R. F. BLACK.

Dr. C. B. Pasley, Acting Surgeon-General, etc.

Dr. Gairdner's cases are reported in the *British Medical Journal*, June 11th, 1887, with the following title : " A Remarkable Experience concerning Leprosy ; involving certain Facts and Statements bearing on the Question—Is Leprosy Communicable through Vaccination?" By W. T. Gairdner, M.D., LL.D., Professor of Medicine in the University of Glasgow :—

"The time seems to have arrived when, without injury or offence to anyone concerned, it is possible to bring under the notice of my medical brethren some facts, and some inferences arising more or less directly out of the facts, in a case which occurred to me some years ago, but which I have found it necessary hitherto to deal with as involving matters of professional confidence not suitable for publication. Even now I shall deem it expedient to frame this mere narrative in such terms as shall not point to any definite locality, or to any recognisable person, among those chiefly concerned ; although, by a formal certificate granted only the other day, I feel, as it were, absolved from the last tie that bound me, even under the most fastidious sense of professional duty, to reticence.

"Six or seven years ago, the parents of a young boy, fairly healthy in appearance, but with a peculiar eruption on the skin, brought him to me, and along with him a letter from a medical gentleman whom I had entirely, or almost entirely, forgotten, but who stated himself to have been a pupil of mine in Edinburgh considerably over twenty years before. It is unnecessary to enlarge on the particulars of this case, further than to state that, after more than one most careful examination, in which I had the assistance of my colleague, Professor M'Call Anderson, we came to the conclusion which we announced to the parents, that the boy was suffering from incipient, but still quite well-marked, leprosy in its exanthematous form ; a diagnosis afterwards amply confirmed. What struck me at the time as most peculiar was, that this case, coming from a well-known endemic seat of leprosy (an island within the tropics) and with a letter involving medical details by a medical practitioner of many years' local experience—sent to me, moreover, for medical opinion and guidance—should not have been more frankly dealt with by a diagnosis announced even to the parents, before they left the island. The father of the child was a sea-captain constantly engaged in long voyages—for the most part between this country and the island alluded to. Both father and mother were Scotch, and there were several other children, all reported as quite healthy, as also were both the parents. Under these circumstances I wrote to the medical man—who in the sequel may be called, for brevity, Doctor X.—simply stating the diagnosis arrived at, and indicating the line of treatment proposed. The parents were informed that it would be best for the child to live in this country, and his mother agreed to remain with him accordingly. And, as they appeared anxious to have every available suggestion and advice, I mentioned the name of Dr. Robert Liveing as having given much attention to the subject, and offered to write to him if they would take the boy to London, as they appeared desirous of doing. Although I wrote to Dr. Liveing, circumstances unknown to me led to a change in their plans, and, instead of going to London, they went to Manchester, where I believe some physician was consulted, but I do not remember who he was. Ultimately, the mother determined for a while to settle in Greenock, and I placed her accordingly in communication with Dr. Wilson of that town, who for some time thereafter remained in medical charge of the case.

"Meanwhile, the course of post brought me in a few months a reply from Dr. X., not only entirely assenting to our diagnosis as communicated to him, but stating that he had been perfectly well aware from the first of the case being one of leprosy, but had deliberately chosen not to affirm the fact or even to allude to it in any way, either in his communications with the parents or in his letter to me. No reason was assigned for this (as it appeared to me) very remarkable reticence; but, as I did not wish to have the credit of having discovered for the first time what a gentleman so much more familiar with the disease might have been supposed to have overlooked, I took means to inform the parents of Dr. X.'s reply, and of his having been all along of the same opinion with regard to the disease as we were.

"After this the matter passed out of my mind, and for several years I neither saw nor heard of this child except accidentally and in a way entirely to confirm first impressions. About three years ago, however, while engaged in lecturing on specific diseases, and among others, briefly, on leprosy, I made an effort to find out something more about this patient. The mother had removed to Greenock, and had brought over the whole family to Helensburgh, where, as I learned, they were visited by Drs. Reid and Sewell, and from the latter I now learned that the poor boy had gone steadily to the worse, and was extremely feeble, covered with sores, and in a most deplorable condition physically, but still receiving every attention and care that constant medical treatment, with the most faithful and loving maternal nursing, could afford to lighten his sufferings. I accordingly proposed, within the next few days, a visit to my old patient, as a matter of satisfaction to myself. Unhappily there was no other apparent object, either as regards diagnosis or treatment, for a visit which was, nevertheless, very gratefully accepted.

"The case was now in the most advanced stage of leprosy, proceeding to mutilation of the extremities, and accompanied not only by external sores, but presumably by internal lesions, which had reduced the patient to the last stage of emaciation. It was on this visit that the curious particulars now to be related were first brought to my knowledge by Dr. Sewell, and afterwards confirmed by the statement of the mother, showing very clearly, though, of course, upon second-hand information to a certain extent, that Dr. X. had a very special reason for his extraordinary reticence in the first instance. Her husband, who in his frequent voyages had opportunities of coming into communication with Dr. X., had remarked to him how very strange it was that, even in writing to a medical man about the case, he had given no hint of his opinion about it. The Doctor's reply to this was, in the end, to the effect that he had kept silence because he did not wish to compromise a boy of his own, whom he (Dr. X.) believed to be a leper, and from whom he believed at the time that the boy he had sent to this country had become infected with the disease. He further explained that he had vaccinated his own boy with virus derived from a native child in a leprous family, and, as I understood (though perhaps not definitely so stated) that leprosy had declared itself in the native child after the vaccination; and, further, that (using his own child as a *vaccinifer*) he had vaccinated our patient directly from him. Before sending the last-named patient away with his parents, he had satisfied his own mind not only that his own boy was leprous, but that he had in this way become the source of the disease to another; but, the disease in his own child being in a very mild form, he was anxious not to disclose its existence. Meanwhile Dr. X. had died; his estate had passed into the hands of trustees; and I was informed that this reputed leper-boy had been, under the instructions of his ather and his guardian, placed and retained at a public-school well known to me in this country, and that the boy was pursuing the usual course of a public-school education, in entire unconsciousness or the disease with which he was supposed to be affected.

"This information, so communicated, placed me in rather a difficult dilemma, namely—was I justified in taking steps to ascertain the truth of the story as regards Dr. X.'s boy, either by personal investigation or, at least, by inquiries conducted so as to result in a well-grounded and scientifically exact opinion as to the facts? And, further, supposing that such opinion should turn out to be that Dr. X.'s boy was a leper, was it a matter of duty on account of others to formally disclose the fact, be the consequences to the boy what they might? It was hardly probable that a boy generally known to be a leper would be retained permanently in any public school in this country, even had it been unquestionably a matter of medical doctrine that such a proceeding was quite safe. On the other hand, the boy was receiving the benefits of an English education at the express wish and on the responsibility of his father and guardian, and without (so far as appeared) any misgivings on the part of anyone. He was an orphan, and in what was to him a foreign land; his remaining under instruction might be, and probably was, a matter of the greatest possible importance to him. To bring him, therefore, even by an indiscreet inquiry, under the ban which in many or most countries still attaches to leprosy was certainly no part of the business of an outsider, and could only be justified at all by an overwhelming sense of duty to others.

"Under these circumstances I thought it well to consult, privately, one or two of those friends in London whom I believed to know most about leprosy, and among others Dr. Liveing, whom

I was able to remind, at this stage, of my previous letter. These friends concurred in assuring me that, in the rather improbable event of their being personally consulted as to the retention of a leper in a public school (it being presumed, of course, that he was physically fit otherwise), they would have no hesitation at all in affirming that the other boys would not be endangered by such proceeding. As I happened to be very well acquainted with one of the medical officers (though not the ordinary medical officer) of the school in question, I communicated these opinions to him, and stated to him at the same time the extraordinary circumstances which had begotten, for me, such a lively interest in the son of Dr. X. In the course of a few days I was informed that an inquiry had been held by the medical staff; that the boy had been sent for and privately examined (though not ostensibly ill in any sense); and that it was, beyond all doubt, considered to be a case of leprosy. The medical authorities decided, however, that under the circumstances it was not their duty to sound the alarm, or in any way to disturb the boy's education.

"From this time onwards (except the death of the first patient soon afterwards) I heard nothing more of these matters till a few weeks ago, when I was asked to see Dr. X.'s son professionally on behalf of the school authorities; and, if so advised, to request Dr. Anderson also to give an opinion as to the present state of health of this young man, who happened at the time to be visiting some friends in Glasgow. It was represented to me that he had maintained, on the whole, fairly good health since I last heard of him through my medical friend, and had not been incapacitated from school work except on account of a contagious eczema which had been prevailing, and with which he had been affected in common with other boys. Apparently, however, the opinion had arisen that his general health was not quite so good, and that, in view of a cutaneous affection of this kind, apparently communicable, existing, it was no longer expedient that he should remain at the school. Indeed, I could not but come to the conclusion that his removal, on public grounds, had been practically settled; and with every desire to soften the blow as much as possible to the poor boy, it was felt to be necessary that his guardian, at least, should receive unequivocal and unbiased testimony as to the actual state of the facts and circumstances under which the decision was arrived at. Under these circumstances I saw and examined this boy, and made a report, along with Dr. Anderson, to the effect that the disease was evidently leprosy, though of a remarkably mild type, as shown by discolorations and cicatrices, and also by large anæsthetic areas on the back of one limb. All breaches of surface, however, and all discharge had ceased at the time of our report, and Dr. Anderson felt still in a position to affirm that no danger to others could occur from the boy's remaining at school. On this last point I did not feel able to give an unqualified assent to my colleague's opinion; but as regards the matters of fact and observation there was no doubt whatever, and our report accordingly on these was substantially as above."

In further explanation Professor Gairdner, in a letter to the *British Medical Journal*, August 8, 1887, says:— "In submitting to you some curious facts and statements which had been brought under my notice as bearing on the above subject ("Leprosy and Vaccination": *British Medical Journal*, June 11, 1887) I was exceedingly careful not to obtrude any opinion of my own. It was clear from the first that the mere statement of such detail would waken up some old controversies, and would perhaps involve very serious practical issues; but these considerations did not appear to me to justify witholding the facts, but rather the public statement of them in as unbiased a form as was possible."

Referring to these cases, Dr. C. Burgoyne Pasley, Acting Surgeon General, Trinidad, observes:—" The fact remains, that an unlucky boy, of undoubted European parentage, acquired a most loathsome disease, and died a miserable death as the result of vaccination, carefully or carelessly performed as the case may have been."—*Papers on Leprosy, Government Printing Office, Trinidad. 1890.*

The same medical authority says :—" If by accident
I draw blood in puncturing the vesicle on the arm of
any child, I invariably reject the child as a vaccinifer,
no matter how healthy it may appear, or how abundant
the supply of lymph may be, fearing to inoculate any
constitutional disease, leprosy, syphilis, etc."—*British
Medical Journal, p. 270, July 30th, 1887.*

A correspondent of the London *Daily Graphic*, writing
from St. Kitts, 5th June, 1890, reports the following
under the title of

"A SAD CASE."

"A sad case occurred here a short time ago, which
shows the danger that arises from the practice of
vaccination in an island where leprosy is treated as
of no account. A few months ago a little girl, the
daughter of the Rev. Mr. ——, a Wesleyan missionary,
who came to the West Indies from England two or
three years before, fell ill. On examination by the
doctors it was found that the poor child had contracted
leprosy. The only probable means of communication
was by inoculation; and thus the parents, endeavouring
to save their daughter from the very remote danger of
small-pox, inoculated her with the horrible poison that
will make her life a living death and herself a loathsome
and repulsive spectacle. Hoping that by returning to
England he might get something done for his daughter,
the missionary resigned his charge and made preparations for his departure. But a new trouble awaited him.
The Royal Mail Company's steamers could not take a
leper as a passenger; but one of Messrs. Scrutton's
vessels agreed to take the sorely distressed family to

England. They got on board, and started on their voyage. But ill luck again attended them, for, while leaving the island, and when nearly opposite their old home, the ship struck on a reef; and, although all on board were saved, the missionary and his family remain here, where they have been so sorely tried."

While in Trinidad, I made inquiries of a highly intelligent merchant, who has resided forty-three years in the West Indies, and has always been much interested in the public health. He says the belief is general in the islands that leprosy is being extensively disseminated by vaccination, and he furnished me with particulars of a number of healthy families where leprosy and other diseases have broken out after vaccination, of others who, in spite of a law enforcing vaccination, have preferred to undergo the worry and penalties of prosecution to the terrible risks of this hideous and incurable malady. In some instances the children infected with leprosy have been sent by their parents to France and England, where, after treatment by some of the most distinguished physicians, they have either succumbed to the disease or returned to die at home; and in one case the mother died of a broken heart on seeing her eldest son come back a complete wreck, loathsome to the sight. All the victims described by my informant were in good circumstances, and none were even sent to the Leper Asylum, where only the poor are interned. He says that had he kept a record he would have been in a position to have given details of very many cases, with all the attending circumstances, and adds, " I have come to the conclusion that we are indebted to vaccination for not only this (leprosy), but many other diseases, especially those of a scrofulous nature, as well as syphilis."

In a communication to the *Lazaretto*, St. Kitts, August 25, 1890, Dr. John Freeland, Government Medical Officer, Antigua, West Indies, says:—" In some of these islands leprosy has no doubt spread for the want of precautions to separate the diseased from the healthy, from poverty, overcrowding, or decomposed food, and from, I fear, the system of arm-to-arm vaccination which now so universally prevails."

The following letters, read before the Royal Vaccination Commission, from Dr. Charles E. Taylor, of St. Thomas, Danish West Indies, secretary and member of the Colonial Council of St. Thomas and St. John, member of the Board of Health, etc., illustrate the difficulties which obstruct the investigation into this momentous subject, owing to the dread which is generally felt by inhabitants of its becoming known that members of their family are tainted with this fearful malady, and probably also to an unwillingness to cast reproach on a prescription so extensively recommended by the profession as vaccination:—

St. Thomas, Danish West Indies
Virgin Islands, Jan. 2 1890.

DEAR SIR,—Referring to your inquiry with reference to the spread of leprosy in the West Indies, I beg to say that it is difficult to obtain testimony with regard to this disease having been conveyed into families either by vaccination or otherwise.

There is such a dread of the hideous fact becoming known, and though parents will talk about such-and-such a case, when it is pushed home to themselves, and their evidence requested for public purposes, even so important as a Royal Commission, they beg to be excused.

My own experience has compelled the conviction that leprosy has on numerous occasions been propagated by the vaccinator's lancet in these islands. Children have been brought to me a year or two after vaccination who have shown unmistakable signs of leprosy, and whose parents assured me that such had never been in their family before. On the other hand, inquiry into the antecedents of the child from whom the lymph had been selected revealed the existence of leprous taint either on the paternal or maternal side.

My own experiences have been confirmed by Dr. Bechtinger, formerly a resident and practising physician here, whose extensive researches entitle his opinion to great weight amongst pathologists.

The belief, also, in the British West Indies as to the conveyance of leprosy in this way is widespread, and forms one of the strongest grounds against compulsory vaccination that I know of.

In view of such a fact, and in face of such a terrible danger, it is my conscientious opinion that every physician should hesitate before subscribing to such a doctrine.

I have the honour to be,
Yours very respectfully,
CHARLES E. TAYLOR, M.D., F.R.G.S., etc.

A later communication from Dr. Charles Taylor is published in the *Public Opinion* of Nov. 27th, 1891. It is dated:—

St. Thomas, Danish West Indies,
October 20.

I have read the report of the evidence given before the Royal Commission on Vaccination in London with much interest, and with regard to the connection between vaccination and leprosy, an experience in these islands of over twenty years enables me to confirm the truth of this terrible indictment. On more than one occasion cases have come before my notice of leprosy in families which could only have been inoculated with the vaccine virus, none of the family having previous to vaccination been afflicted with this malady. Leading dermatologists in all parts of the world, and the most experienced physicians in the West Indies, are of the opinion that leprosy is spread most readily by means of inoculation, either through a wound or an abraded surface, and still more readily by puncturing contaminated vaccine virus into the arms of healthy persons. The reports of the medical officers of health and physicians to the leper asylums in the West Indies show that leprosy, which thirty years ago was stationary or subsiding, has increased. This, I have every reason to believe, and it is also the opinion of other competent medical men, is coincident with the introduction and spread of vaccination, for there are a number of islands where the disease was almost unknown previous to its inoculation in this way. Were it not for the reluctance which all physicians, have to expose families tainted with leprosy, they could give evidence as startling as the cases mentioned by John D. Hillis, of British Guiana; Dr. Bechtinger, formerly of St. Thomas; Dr. R. Hall Bakewell; and Dr. Black, of Trinidad. The possibility of spreading such a dire disease by means of the lancet is one too grave to be longer disregarded, and, it is needless to say, a serious matter for these islands, the most lovely in the world, where children, whose parents may be the most healthy, are liable to leprosy through arm-to-arm compulsory vaccination. May I venture to hope that the English Press will have the humanity and courage to speak out and compel colonial authorities to withdraw the vaccination enactments, which on these grounds alone are so dangerous to ourselves and our families.

CHARLES E. TAYLOR, M.D.

So great is the dread of the invaccination of leprosy and syphilis, that when I visited the island of Granada in January, 1889, a gentleman connected with one of the public institutions of the island told me that he had two unvaccinated children, and, that rather than incur the risk of invaccinated leprosy or syphilis, he had sent them with his wife to a place of refuge in the mountains. He was not sure, he said, whether this *ruse* would succeed, as the authorities were very sharp.

When I visited Barbados in January, 1889, Mr. E. Racker, the proprietor of the *Agricultural Reporter*, Bridgetown, informed me of a case of leprosy communicated by vaccination, which he had personally investigated. Mr. Racker was intimately acquainted with the father, a member of the Legislative Council, and on one occasion, when visiting his house, noticed that his friend's youngest child was afflicted with leprosy. The father said it was due to vaccination with lymph taken from a child subsequently discovered to be leprous. Though he believed in the benefits of Jenner's discovery, he declared that there should be no more vaccination in his family. I may observe that so widespread is the belief in Barbados that leprosy and syphilis are communicated by vaccination, that every attempt to make it compulsory has been defeated. Nevertheless the advantage of vaccination is believed in by several men with whom I conversed, and there is a considerable amount of vaccination practised, to which I attribute no small share in the admitted augmentation of this disease before referred to. The hideous risk attending the practice of vaccination is illustrated by the following letter of Mr. Racker, who writes 2nd May, 1890:—" I know all about the case reported in the *British Medical Journal* by Dr. W. T. Gairdner. I am one of the executors to the will of Dr. J. C., but I had no idea that the boy was suffering from leprosy until I got a letter from the head master of Dollar, enclosing a letter from Dr. Gairdner.

" I think I told you how I once consented to have my children vaccinated, and how at the last moment I changed my mind, and would not allow them to be operated on. Well, that boy, Dr. J. C.'s son, was the one from whom they were to be vaccinated. What a narrow escape I had!"

Mr. Alexander Henry, Vice-Chairman of the Council of the British and West Indian Alliance, and formerly editor of the *St. Kitts Gazette*, who has resided some years in the West Indies, and has devoted much attention to the spread and causation of leprosy, writing 12th June, 1890, says:—" A medical officer of health cautiously admitted to me that leprosy was contracted by means of careless vaccination. Now, careless vaccination means vaccination from arm to arm, which is almost universal in these islands. I do not believe there is a doctor of any standing in the West Indies who would deny that leprosy can be inoculated. It is admitted that owing to the slow incubation of the disease it is difficult to distinguish a leper; and when you take into account that medical officers are constantly complaining to the Government 'that they cannot get a supply of calf-lymph,' and add to this the indiscriminate and careless yet vigorous manner in which they carry out the vaccination laws upon an ignorant and simple people, who have no means of asserting themselves, I think we may safely conclude there is a high probability that leprosy is spread by vaccination."

When or how leprosy entered this island is perhaps unknown. Certain it is, however, that the dread disease has recently been advancing by leaps and bounds. In proof of its increase here I quote the following passage from a recent quarterly report of Dr. Alfred Boon, one of the Government medical officers and acting analysts of Government statistics. Dr. Boon says:—

"The opinion that it is, as many hold, increasing in the island, is supported by the fact that in the four years 1885 to 1888 thirty-four such deaths were registered."

In a later communication received by me from Mr. Henry, dated October 8, 1891, he declares his conviction that vaccination is one of the chief factors in the spread of this fearful disease in the West Indies. After referring to the small-pox epidemic in Martinique, in 1887, and the consequent extensive re-vaccination propagated in St. Kitts, the writer proceeds :—" But, you may ask, how did that affect the labouring class, who are known to have a strong prejudice against vaccination, and among whom the disease of leprosy is most common? This is the incident to which I wish to draw your attention. There is in force in St. Kitts (and in most of the other islands) a law to compel all boatmen and porters to take out a yearly license. Now, it so happened that, at the time when the boatmen and porters required to renew their yearly badge, the small-pox was raging in Martinique, with, of course, the usual panic in St. Kitts. The boatmen and porters were informed by the Inspector of Police, or by his authority, that no one could have his license renewed unless he had been re-vaccinated that year. I need not point out the unspeakable dangers which such a system of indiscriminate—almost reckless —arm-to-arm vaccination exposed adults of the class among whom leprosy is rampant. In considering the medical aspect of the practice, you must bear in mind that the defence of the arm-to-arm vaccination by the medical officers is their admission that it is almost impossible to obtain pure calf lymph in the islands. Can you wonder, then, that there should exist a strong prejudice and an unspeakable dread of vaccination among the lower classes in the West Indies?" In support of this contention, that the majority of the West Indian islanders fear the practice of vaccination, Mr. Henry adduces the evidence of the medical officers themselves :—" Dr. Gavin Milroy, in his ' Report on Leprosy and Yaws in the West Indies' (House of Commons Command Papers, c. 729) states on pages 32, 33, ' In the frequent conversations which I subsequently had with many of these gentlemen (the medical officers in the West Indies) I learned the fact that the European and most of the higher creole families were always extremely anxious about the source of the lymph to be used in the vaccination of their children, from the dread of a leprous taint being thus acquired. None of my informants appeared to partake of this belief themselves, but all recognised the propriety of avoiding the use of lymph from children of families known or believed to be afflicted, especially as infants themselves rarely, if ever, exhibit any outward manifestations of the malady.' " Dr. Milroy says that the Vaccinator-General, Dr. Bakewell, seemed to give countenance to the *popular belief* as to the transmissibility of leprosy by vaccination."

Dr. de Verteuil, of Trinidad, in replying to the views of Dr. Bakewell, says (p. 34) that "it is necessary to take certain precautions. The vaccinifer should be healthy, and born of parents free from any syphilitic or leprous taint, and, as hereditary syphilis generally manifests itself after the age of four or five months, it would be as well to choose as vaccinifers children of five or six months and above." The only comment I need make on the "defence" of Dr. de Verteuil's is to point out that it is well nigh impossible to take the necessary and certain precautions to know if the vaccinifers are *leprous*.

Dr. Bowerbank, of Jamaica, in pp. 34, 35, says:—"I have frequently heard of a case of leprosy occurring in a family, alleged to have been the result of vaccination. I know of two instances in one family (a Jewish one), in which the parents and friends are thoroughly convinced in their own minds that such was the case. In Barbados we find a strong prejudice against vaccination, for Dr. Browne writes:—'It has been a general rule not to vaccinate from the *apparently unhealthy*, or those of leprous taint, not so much from any opinion founded on fact of the possibility of conveying the disease, as *from respect to the general prejudice prevailing*. For the Leeward Islands, I refer you to a letter to the editor of the *St. Kitts Lazaretto*, 23rd August, 1890, by Dr. Freeland, Government Medical Officer of Antigua:—'In some of these islands, leprosy had no doubt spread from the want of precautions to separate the diseased from the healthy, from poverty, from overcrowding, or from decomposed food, *and from, I fear, the system of arm-to-arm vaccination, which now so universally prevails.*'"

Mr. Henry says that the only point on which medical men seem to be unanimous regarding leprosy is that, whether it is contagious or not, it is inoculable; and, after citing various medical testimonies, showing the connection between leprosy and vaccination, he observes :—" If ever a case was proved it is this one that there is a universal belief in the West Indies that leprosy is spread by means of vaccination. And I, for one, place far more reliance upon a popular belief of this kind, and especially upon this subject, than I place upon medical testimony, because the neighbours of a leper are far more likely to know when and how he contracted the disease : they know his pedigree, family history, and the whole condition of his life and its surroundings, whereas the medical officer simply pays a casual visit, and often does not hear or know of the leper till the disease manifests itself to such an extent that the leper becomes a burden and a danger to his relations and his neighbours.

"But what was the answer of the College of Physicians to the inquiry of the Chief Secretary of State for the Colonies, as to 'whether there was any ground for the belief that leprosy was spread by vaccination (in the West Indies)?' On page 86, same Report, in their answers and advice to Lord Kimberley:—'The College of Physicians feel they cannot press too strongly on your Lordship the importance of enforcing the practice of vaccination for the protection of those who are too ignorant to protect themselves.' As the people become a little wiser, they will be able to protect themselves from the prejudice of the College of Physicians."

SANDWICH ISLANDS.

Leprosy is a disease of relatively slow incubation. Children pronounced perfectly healthy, and represented as approved subjects both for vaccination and for vaccinifers, may, in a few months, exhibit the unmistakable signs of leprosy. In a report by Dr. Edward Arning, dated Honolulu, H.I., November 14, 1885, presented to the president and members of the Board of Health, p. 52, this distinguished observer says:—"The next point touches the vaccination question, with which I have dealt at length under the heading of experimental work. I would further urge that the medical examinations of school children, which have led to the elimination of quite a number of cases, should be kept up regularly and carefully. As an instance of their necessity, I may quote a case which has quite recently come under my observation. A little girl (native) belonging to one of our large schools passed my close examination a year and a half ago as healthy, but now presents initial symptoms of leprosy. We must not rely on general healthy appearance in these examinations, and on a furtive glance at hands and arms. I have found unmistakable marks of leprosy on the back of a child that held a recent health certificate. Moreover, we shall have to extend our examinations even to the very young children, in spite of Dr. Fitch's assertion that leprosy does not make its appearance before the period of second dentition. I have seen a child with clear signs of leprosy at three and a half years of age, and know of another boy who was a marked case at four years old."

This eminent bacteriologist, in a letter before me, dated September 6, 1889, says:—" During my stay on the Hawaiian Islands for the bacteriological study of leprosy, I was naturally drawn to a scrutiny of the question whether leprosy is transmissible, and had been there transmitted by vaccination; all the more so as there is a general opinion prevailing on these islands that the unusually rapid spread of the disease about thirty years ago may possibly be attributed to the great amount of indiscriminate vaccination carried on about that period. And there is no mistake about the actual synchronicity of the spread of vaccination and of leprosy in the Hawaiian Islands; but many a mistake is possible as to the real causal relation between the two.

"I could trace the first authenticated cases of leprosy back to about 1830, but the terrible spread all over the islands did not take place until very nearly thirty years later, at a time when an epidemic of small-pox had given rise to very general and very careless vaccinations throughout the group. . . . . I attach far more importance to an instance of an increase of leprosy soon after vaccination on a much smaller scale, and during a much more recent period than the above. I have it on good authority that a very remarkable new crop of leprosy cases sprang up at Lahaina, on the island of Maui, about a year after most careless vaccination had been practised there."

The impossibility of detecting leprosy in its early stages is a matter of common notoriety amongst physicians, so that many who believe in the prophylaxis of vaccination refuse to incur the terrible risks involved by its practice in leprous countries.

Dr. E. Kaurin, in "Notes on the Etiology of Leprosy," in the *Lancet*, January 25th, 1890, observes :—" We must bear in mind that the duration of the disease is much longer than is generally supposed. The patients, as a rule, take no account of the long prodromal stage, marked by indefinite subjective phenomena and temporary affections of the skin. The physicians themselves often overlook the early symptoms of the disease; indeed, it is sometimes impossible to form a diagnosis in the early stages. The whole period of the disease, from the onset to the time when distinct signs are noticeable, amounts to at least from three to four years."

Mr. H. A. Acworth, writing to me from the Municipal Commissioner's Office, Bombay, July 29th, 1891, says:— " I have plenty of lepers in my hospital here who could not be identified as such, unless they were completely stripped and examined by a trained eye."

Dr. Mitra, in his Report on Leprosy in Kashmir, refers to the recorded statistics of lepers, and says that the enumerators " might have overlooked many cases in the incipient stages," whereupon the Lahore *Civil and Military Gazette* calls attention to the consequent dangers of communicating organic or other diseases by means of vaccination.

Dr. E. Arning says:—" When in Hawaii I attended a German boy, aged 12, who suffered from leprosy, from whom, when he was seven years old, several white families had been vaccinated."—*Journal of the Leprosy Investigation Committee, February, 1891, p. 131.*

In the report of Dr. Webb, medical inspector, on the condition of the schools, April 1st, 1886, are recorded five cases of incipient leprosy in the Royal School, and two in the Fort School, Honolulu.

Dr. F. B. Sutliff, of Sacramento, California, who has studied the disease as Government physician on the Island of Maui, says :—" I very seldom visited a school without excluding some (children), while the spots just beginning to show in others made it only too probable that they would not long be doubtful cases. It did not seem to me a difficult task to read the fate of Hawaii in the little dark faces that looked up from their books."— *Occidental Medical Times, April, 1889.*

In Dr. Arning's report, dated Honolulu, Nov. 14, 1885, p. 14, we read :—" Closely allied to inoculation is the subject of vaccination. You are doubtless aware of the very prevalent opinion among medical men that the unusually rapid spread of the disease may possibly be attributed to the great amount of indiscriminate vaccination which has been carried on in these islands. If my information is correct, unquestionably new centres of leprosy have developed after vaccination was practised, and several old inhabitants have told me how they themselves used no precautions whatever in vaccinating during a small-pox scare.

" To bring some light on this moot point I vaccinated a number of lepers. The vaccination only took in three cases, one tubercular and two anæsthetic. Both the lymph and crust of the tubercular case contained the *bacillus lepræ;* in the anæsthetic cases I could not detect it."

Dr. Arthur Mouritz, then Resident Physician and Medical Superintendent of the Leper Settlement, Molokai, in his Report for 1886 to the Board of Health, after alluding to contagious and hereditary predisposition, says:—" The third cause to which I attach some importance, and which has undoubtedly spread the disease, is vaccination. I can bring forward no case personally, but I have reliable hearsay evidence that after the operation of vaccination had been performed on several white children, they manifested signs of leprosy, and finally developed the disease. Evidence on this same point is put forward by Sir Ranald Martin in India, and by Professor H. G. Piffard, of New York, both reliable authorities."

An Appendix to the "Report on Leprosy" addressed to the Legislative Assembly of Hawaii, in 1886, is an interesting account of Queen Kapiolani's visit to the Leper Settlement at Molokai, by the Princess (now Queen) Liliuokalani, in July, 1884. Amongst other incidents, the Princess refers to an interview with one, Kehikapau, in the presence of several persons. Kehikapau called the Princess's attention to the circumstance of his having contracted the disease from vaccination. He also mentioned that, through the same agency, all his schoolmates had died of the disease, induced in this way.

According to the Report of Surgeon J. R. Tryon, of the United States Navy, leprosy has spread "from year to year in Hawaii, and has increased to a marked degree since the indiscriminate and careless vaccination practised during the severe epidemic of small-pox in 1853."—*Medical and Surgical Memoirs, 1887, vol. 2, p. 1252, by Dr. Joseph Jones, President of the State Board of Health, Louisiana.*

In a summary of reports furnished by foreign Governments to His Hawaiian Majesty's authorities as to the prevalence of leprosy in India and in other countries, and as to the measures adopted for the social and medical treatment of persons afflicted with the disease (Honolulu, 1886), I find the following extracts, p. 238 and 239, from the *New Orleans Medical and Surgical Journal, April, 1880.*

After referring to the relation of leprosy with syphilis in the Hawaiian Islands, the author says:—"Vaccination was also inquired into. Alarmed by an invasion of small-pox in 1853, a general vaccination of the whole population was ordered, and physicians being at that time very few on the islands, non-professionals aided in the work.

"It is charged by some that, as a natural result of the labours of the heterogeneous force so appointed, not only syphilis, but also leprosy, was greatly increased. In my last circuit trip in my district, I found very few adults who had never been vaccinated. This involves the question of inoculability—in my opinion the main, if not the only, means of propagation other than inheritance—that is, like syphilis, it depends for its propagation upon the direct introduction of virus into the blood."

The fact that an increase in the practice of vaccination in leprous countries is often accompanied by an increase in the dissemination of leprosy is shown by the following evidence:—

Mr. George C. Potter the Secretary to the Honolulu Board of Health, writing to me on behalf of the President of the Board of Health, Dr. J. H. Kimball, in a letter dated Honolulu, H.I., June 1st, 1890, says:—"It is an opinion among the laity and some of the profession that the extensive arm-to-arm vaccination that was practised in the years 1852 and 1868 during small-pox epidemics was a prolific cause of the spread of leprosy." In a report of the Board of Health to the Legislative Assembly of 1886, by the then president, Mr. Walter M. Gibson, I read, p. 35, "There are two more causes which, in my judgment, have had a great effect in the propagation of leprosy, or diseases closely allied to it, although, medically, it be a disease *sui generis*. The first was the ignorance of some of the early and unqualified medical practitioners who were permitted to spread disease broadcast, and to do irretrievable injury before retribution overtook them; but the second and chief cause was the indiscriminate and, to my mind, careless vaccination that began about 1868."

Dr. H. S. Orme, President of the State Board of Health, California, in an able memoir on " Leprosy: its Extent and Control," says:—" There can be no doubt that the lowering of the vital stamina of the race by the great prevalence of syphilis (at Hawaii) prepared them for the inroads of any disease that might threaten. During this period small-pox also scourged the people, and in 1868 there began a general vaccination, in which virus was taken indiscriminately from human subjects. This reckless practice doubtless contributed greatly to the spread of both syphilis and leprosy."

In an article on " Personal Observations of Leprosy in Mexico and the Sandwich Islands," in the *New York Medical Journal* for July 17, 1889, the author, Dr. Prince A. Morrow, M.A., says:—" Vaccination is believed by the natives, as well as by many intelligent physicians, to be a potent agent in the rapid diffusion of leprosy through the islands. It must be remembered that until recently vaccination was performed by unskilful persons, human virus was used, and no distinction was made between a healthy person and a leper as a vaccinifer. The fact is incontestable that, after the general vaccination of the natives, numerous leprous centres developed in various parts of the islands where the disease had previously been unknown. Arning demonstrated the plentiful presence of *bacilli* in the lymph and crusts of vaccine pustules in lepers."

In the *Archiv für Dermatologie und Syphilidologie* for January, 1891, we read that at the Berlin Medical Congress of 1890 Dr. Arning read a paper on "The Transmission of Leprosy," wherein he scouts the idea of hereditary transmission. He proceeds to ask the question raised by the late Dr. Hillebrand, "Has leprosy been spread in the Hawaiian Islands by means of universal vaccination?" And he declares:—

"There can be no doubt as regards the synchronousness of the diffusion of leprosy and the introduction of vaccination into the Hawaiian Islands.

"I am able to state—having excellent authority for so doing, though, unfortunately, no statistics—that a very remarkable accumulation of fresh leprosy cases took place in 1871-72, in a place called Lahaina, on the Island of Maui. This happened about one year after a universal arm-to-arm vaccination, which had been most carelessly performed. About fifty to sixty cases occurred suddenly in this locality, which up to that time had been comparatively free from the disease."

Dr. Arning concludes:—" Arm-to-arm vaccination should be prohibited in countries in which leprosy abounds."

When, during my visit to Honolulu in October, 1890, it became known that I was seeking information concerning the spread of leprosy, in the interest of the public health, and not to support any medical theory or foregone conclusion, several gentlemen who had devoted much attention to these subjects called upon me, and others gave me introductions to those who were conversant with them. The President of the Board of Health, the Hon. David Dayton, and the Secretary, Mr. Potter, afforded me valuable assistance, and supplied me with copies of their various official reports relating to the introduction and dissemination of leprosy and the methods adopted for dealing with the scourge, which, with other European diseases, bids fair, unless arrested, to destroy the entire native race. They incidentally showed me copies, cut from the Press, of my own communications on the subject of leprosy and vaccination in the West Indies, thus illustrating their desire to inform themselves upon the subject by evidence from all quarters of the globe.

I have, in this monograph, made free use of the facts and testimonies contained in these important official documents, and I take this opportunity of expressing my thanks for the courtesy and information furnished by the gentlemen connected with the Board of Health, resident physicians, the Executive Officers at Molokai and Kalihi, and members of both Houses of the Legislature.

According to all the evidence which I have been able to obtain, leprosy was unknown in the Sandwich Islands until many years after the advent of Europeans and Americans, who introduced vaccination; and there is no aboriginal word in the Hawaiian language for this disease. Mr. Dayton, President of the Board of

Health, says that the natives, having no words of their own, used the Chinese words *mai pake?*—" what is this disease?"

In Captain Cook's time these islands were supposed to contain a population of 400,000; at the present time they do not number more than 40,000, and are rapidly diminishing. In all quarters, both native and European, lay and medical, among members of both Houses of the Legislature, I found the belief all but universal that leprosy was considered to be communicable, and that the propagation of the disease during the last twenty-three years was largely due to vaccination.

One medical authority told me that he had no doubt that the disease was inoculable and spread by vaccination, but he did not think it would be prudent to disclose the fact amongst the natives, as he would not be responsible for what they would do. He expressed his own convictions freely on the subject, which are confirmed by Dr. Edward Arning's inoculation experiment on the condemned convict Keanu, of whom he gave me a photograph, showing the development of the tubercular form of this disease. No other intelligent resident shared the fear of such an exposure, the incriminating facts having already been acknowledged both in the official reports and in various communications by medical practitioners to American medical journals. On the contrary, when it became known that I was there not to institute experiments but to collect facts in the interest of those afflicted people, I was urged by influential citizens, and particularly by members of the Legislature, to do what was possible to make known the evils under which they suffered, and to bring an enlightened public opinion to aid them in putting an end to a mistaken medical procedure, which had led to such disastrous results. One of the gentlemen who besought my intervention was Mr. J. Kalua Kahookano, a barrister-at-law, and representative of North Kahala, Island of Hawaii (the largest in the Archipelago), who introduced a bill (July, 1890) in the Hawaiian Legislative Assembly to repeal the vaccination laws. This bill was supported by a petition from Mr. Kahookano's constituents, showing how leprosy, syphilis, and other diseases had been scattered broadcast in these islands by means of the vaccinator's lancet, and new centres of the diseases thus established. The truth of this fearful indictment is now admitted by several medical practitioners of high standing, who have visited the islands to study the cause of the rapid spread of this destructive malady.

Amongst other old residents who kindly volunteered information was Mr. H. G. Crabbe, a member of the Upper House of Legislature, who had resided in the islands for forty-three years, and had always taken an active interest in the public health. He said it was time that the true facts concerning the propagation of this disease in Hawaii were made known, as the people were being decimated by leprosy conveyed to the blood by vaccinators. Mr. Crabbe detailed various facts in proof of this serious charge, and expressed the utmost anxiety that the truth should be made manifest in countries where public opinion was a potent factor for the well-being of the community. He had met with many cases of leprosy clearly traceable to vaccination in the islands, and the facts were admitted by those who, like himself, had taken the trouble to investigate them; but the evil was being perpetuated by those who ought to know better, partly through apathy on the part of the authorities, and partly through the ignorance of the natives, who accepted whatever kind of vaccine virus was most handy when wanted for use. Mr. Crabbe was a believer in vaccination as a prophylactic against small-pox, but considered that, as at present administered, it was a most cruel and mischievous infliction upon a confiding population.

In a statement handed to me by Mr. Crabbe, dated Honolulu, Hawaii, Oct. 22, 1890, the writer says:—" In the year 1866 there was an indiscriminate vaccination of young and old amongst the natives; this vaccination was compulsory, but, thank heaven, I did not allow my children to be vaccinated with the common herd. From that time the leprosy cases became more frequent; many natives who had previously been healthy were afflicted with leprosy. The spread of the loathsome disease was more pronounced in the years following this indiscriminate vaccination. I can recall one case in particular of a native girl by the name of Kapeka, as healthy and as nice looking as 'tis possible for a native to be, who was forced to be vaccinated in the year 1866. In the year 1871 or 1872 Kapeka was sent up to the Molokai Leper Asylum by order of Dr. Robert M'Kibbin. In this particular case I have always contended that this girl was inoculated with the germs, not only of leprosy, but also of syphilis, from this fact. Long before she exhibited any signs of leprosy (some three years), if she hurt herself, so as to make an abrasion of the skin, the place would inflame and suppurate, and would take a long time in healing, presenting an appearance like a syphilitic sore. She died at the Leper Settlement."

MORE STRINGENT VACCINATION RECOMMENDED.

The appropriation by the Government for the Molokai Lazaretto is £10,000 a year—a large sum for a poor and comparatively thinly populated country. During the past twenty-five years, about a million and a quarter dollars have been expended by the Hawaiian Government in making provision to benefit these afflicted people. It is, however, melancholy to reflect that one source of the dissemination of the disease, viz.—vaccination, is still permitted, and, there is reason to believe, is encouraged both by the misguided Government and, notably, by certain official members of the medical profession. The latter have inherited such a deeply-rooted prejudice in favour of the merits of the Jennerian practice as to blind them to its destructive potency in countries where leprosy is endemic. The great majority of medical practitioners take their views from the medical journals, which most unfairly refuse to give their readers the adverse side of the Jennerian practice.

Notwithstanding the evidence of the disastrous results of vaccination in spreading and establishing new centres of leprosy, we are told in the "Biennial Report of the President of the Board of Health to the Hawaiian Legislature for 1888" (in which the depopulation of the islands, and the spread of leprosy, is frequently referred to) that "the work of vaccination has been pushed with vigour;" and, "The Board would recommend the passage of a more stringent law, imposing heavier penalties and giving the vaccination authorities all necessary authority."

In his remarkable work, "Traité Pratique et Théoretique de la Lèpre," p. 306, Professor Henri Leloir refers to the introduction and progressive increase of leprosy in the Sandwich Islands, and states that vaccination, compulsory and *en masse*, contributed to the spread of the disease.

Coincident with the activity with which vaccination has been extended in Hawaii, there have been several very severe outbreaks of small-pox. In the year 1881, according to official reports, 500 persons died of small-pox, a large majority of whom had been vaccinated. Each of these epidemics has been accompanied by a more stringent enforcement of vaccination, and has been followed by the development of new centres of leprosy and the more rapid spread of this destructive scourge.

BRITISH GUIANA.

In the pursuit of my inquiries in British Guiana in 1888-89, the public librarian, Mr. Rodway, expressed himself much interested in the subject of leprosy, which, considering the remarkable increase of the disease in that colony, has been neglected of late years. He called my attention to a work entitled "Leprosy in British Guiana," by a careful scientific observer, Dr. John D. Hillis, formerly Superintendent of the Leper Asylum in that country. Mr. Rodway said this was regarded as the standard work on the subject. From it I extract the following particulars of invaccinated cases:—

Case IV., p. 30.—*Confirmed Tubercular Lepra, supposed to have been contracted by Vaccination.*—Joseph Francis C——, a fair Portuguese, born in Demerara, now aged twenty years. His parents are alive and healthy. He has been suffering for the last ten years from tuberculated lepra. He has a sister, aged eighteen years, at present (1879) an inmate of the asylum, suffering from the same form of leprosy. They were both admitted on July 30th, 1877, from Murray Street, Georgetown. They have three sisters and one brother, who are alive and well. Our patient, J. F. C——, and his sister were vaccinated with lymph obtained from a member of a Portuguese family,[1] in whom leprosy was afterwards found to exist. They were the only members of the C—— family vaccinated with this lymph. Within 18 months of the performance of the operation by Dr. ——, a reddish brown spot appeared on the inner side of the right thigh, preceded, it is stated, by some constitutional disturbance. This spot was raised and tender, accompanied by profuse sweating all over the body, and remained for some time. Subsequently other spots made their appearance on the right buttock (which disappeared shortly after), between the shoulders, and on each cheek. They were all ushered in by more or less well-marked febrile symptoms. A red patch next appeared on the forehead, and epistaxis set in, periodically occurring to this day. Tubercules then made their appearance on the face, the other patches continuing to increase in thickness and roughness, and forming tubercular infiltration."

"The latter was removed by gurjun oil, under which treatment many of the symptoms were ameliorated." State and condition on November 30th, 1879:—

"He has a light brown irregular patch on the front of his chest; this had been larger, thicker, and mahogany-coloured, and has evidently undergone partial absorption. There is a patch of tubercular infiltration on the back of

---

[1] "It is within the knowledge of Dr. Manget, surgeon-general, and the author, that this family are at present afflicted with tuberculated lepra."

the arms, and at the back of the elbows. The fingers are swollen, shining, and dark-looking, a solitary tubercle forming on the back of the hand. The swollen condition of the fingers and hands is very characteristic. There are two tubercles on each cheek the size of marbles; the lobes of the ears are thickened, and a tubercle is forming on the upper lip.

"There is no appearance of hair growing on the face. There are reddish-brown discolourations on the front and back of the legs. There are a few small scattered tubercles on the dorsum of the feet, and the lower parts of the legs are swollen and hard to the touch. There are tubercles on the scrotum, an ulcer on the leg where a tubercle has ulcerated, and the larger tubercles are slightly anæsthetic. This young man is one of the carpenters of the Institution; he is in hopes the treatment now being adopted may arrest the disease, which is, however, making slow but sure progress."

On page 191 of Dr. Hillis's book, a passage is cited from the work of Dr. Vandyke Carter (p. 178), one of the greatest authorities on the disease, in which vaccination is instanced as one mode of communicating leprosy. On page 192, we read:—" The subject of leprosy was brought forward at the stated meeting of the New York Academy of Medicine, January 20, 1881, in a communication by Dr. H. G. Piffard, in which the author, who is not himself a believer in the contagiousness of leprosy, states:—' A review of the evidence bearing on the contagiousness of leprosy led the speaker to the conclusion that this disease, like syphilis, is not contagious by ordinary contact, *but it may be transmitted by the blood and secretions. Vaccination may transmit it.*'

A case in the speaker's own experience was cited in proof of this."—*The Medical Record, February 19, 1881, p. 212.*

Sir Ranald Martin states:—"The dangers to Europeans arise chiefly from vaccination and from wet-nursing. I felt that very early in my career in India, and I took the precautions which are here recorded. I saw an English lady last year in a horrible condition (she said) from having been vaccinated from a leprous child."— *Leprosy in British Guiana, by Dr. Hillis, p. 182.*

On page 208 we read :—"I have already given some cases in which there could be no reasonable doubt but that the disease was produced by vaccination with tainted lymph. Those of the brother and sister mentioned are conclusive on the point, and we have the testimony in favour of this mode of propagation from such men as Tilbury Fox and Erasmus Wilson. I will therefore conclude this chapter with a case from the work of a recent writer, Dr. Piffard, of New York":—

"Case III.—William T——, aged 25 years, was admitted into Bell Hospital in May, 1864. He was of English parentage, but was born and passed his early life in British Guiana. After vaccination, performed when young, his arm became greatly swollen and inflamed, and large sloughs separated. Investigation revealed the fact that the vaccine virus had been taken from a negro whose mother was a leper. At the age of seven years some brownish spots appeared upon his back and arms; and at the age of eleven a blister formed on the palm of the right hand, followed by permanent contraction of the flexor tendons. A few months later he felt a tingling sensation around the nail of the right index finger, followed by a line of suppuration and loss of the nail. The finger soon healed, but the same morbid process separated itself in the other fingers of the same hand. After a few months, according to his statement, the skin of the distal phalanges split, and the flesh shrank away from the bones, leaving them exposed. The bones separated at the joints, and the stumps healed. These various processes occupied eighteen months or two years.

"The disease then affected the distal phalanges of the left hand in the same manner. After this it attacked the right foot, and a slough formed over the lower part of the instep. The great toe then became swollen, the skin split, and its distal bone separated; then, without much regularity, the remaining phalangal bones o fingers and toes necrosed and came away."—*Diseases of the Skin, p. 209.*

"On examination," adds Dr. Hillis, "the patient was found to have *maculæ* or other spots, and anæsthesia of the parts affected."

On the 31st March, 1890, I wrote Dr. Hillis, referring to the cases of invaccinated leprosy mentioned in his book, and asked if he could furnish me with any other facts, and whether he would be willing to give evidence before the Royal Commission. To this I received the following reply :—

134 Leinster Road,
Dublin, 2nd April, 1890.

DEAR SIR,—Yours to hand. I have no further reliable evidence as to the transmission of leprosy by vaccination than that contained in my book. These cases, however, may be relied on. I got the particulars from the medical man who performed the vaccination in question, and the parents of the children. I enclose a reprint from Timehiri.—I am, Yours faithfully,

JOHN D. HILLIS.

Wm. Tebb, Esq.

The British Medical Journal of November 5, 1887, contains a letter from Dr. Hillis, repeating his conviction as to the communicability of leprosy by means of vaccination. Dr. Hillis says he has had more than twenty years' experience of the disease (leprosy), and half of this time he was superintendent of the largest leper asylum in the West Indies.

In this remarkable work, Dr. Hillis quotes Dr. George Hoggan, whose testimony I have adduced, as an authority on the subject of leprosy, and few persons have had greater opportunities of studying the pathology of this disease.

In a recently published work by the late Archdeacon Wright, entitled "Leprosy an Imperial Danger" (Churchill, 1889) the writer, p. 85, reluctantly admits the danger of leprous vaccination. "Much, very much," he says, "seems to imply that leprosy can be communicated by inoculation, and is communicated by vaccination."

Dr. C. F. Castor, the Medical Superintendent of the Leper Asylum, Mahaica, British Guiana, in his report for the year 1887, p. 43, says:—"Another manner in which the disease (leprosy) may be produced in the healthy with no taint is vaccination. This seems a most probable means of communicating the disease, nor can there be any doubt, I fancy, after reading the admirably recorded case by Professor Gairdner, of Glasgow, in the *British Medical Journal* of the 11th June, 1887. Dr. Rake, of Trinidad, disputes the obvious conclusion of the professor, and marshals a number of facts that do not in any way, to my mind, overthrow the fact that in that case vaccination was the cause of introducing the disease in the child."

Again, in paragraph 86 of the same report, Dr. Castor says:—"I have noted these points because I consider they are important, and as needlessly obscuring a palpable fact which should be made known far and wide in countries where leprosy is endemic and widespread, as with us, that there is every certainty of inoculation through vaccination."[1]

The dread of communicating leprosy at Georgetown by means of vaccination is very general, and, as a consequence, the vaccination laws are, to a large extent, inoperative. Dr. Robert Grieve, Surgeon-General for British Guiana, in his report for 1887, referring to vaccination, p. 7, says:—"In the beginning of the year vaccination, which had been carried on energetically in

---

[1] Since the above was written Dr. Castor has given evidence before the Royal Commission on Vaccination (December 2nd, 1891), and in some particulars modified this opinion as to the danger of communicating leprosy by means of vaccination. In an article in No. 4 of the "Journal," December, 1891, in reply to my evidence on this subject before the Royal Commission, p. 4, Dr. Castor observes:—"The opinion expressed that vaccination from a tainted source will produce the disease is, I believe, a true one." Surgeon Brunt, R.N., testified, March 2nd, 1892, before the Royal Commission, to actual cases of invaccinated leprosy within his own experience. Positive evidence and a body of unimpeachable facts cannot be set aside by wavering and contradictory statements like those of Dr. Castor.

the latter part of 1886 in Georgetown, came practically to an end, owing to the unwillingness of the people to bring their children for the purpose."

On inquiry from both medical practitioners and intelligent residents, I found that this objection was mainly due to a wholesome dread of infection of leprosy, when the vaccination was performed with arm-to-arm virus, and of syphilis with various cutaneous eruptions, when imported lymph was used.

In the "Report on Leprosy and Yaws in the West Indies," by Dr. Gavin Milroy (House of Commons papers, C 729), the Surgeon-General of British Guiana says (p. 33):—"As far as my own opinion goes, I am inclined to believe in the possibility of such communicability—*i.e.*, of leprosy by means of vaccination. There is no objection in Guiana to vaccination, if the parties know the children from whom the lymph is to be taken." But against this there is the testimony of Drs. Hackett, Watkins, Stevenson, and Allison, who state that vaccination has hitherto been practised to a very limited extent among the lower classes in Guiana. Dr. Allison says that "this is principally, if not entirely, due to the difficulty of obtaining vaccine lymph;" while Dr. Stevenson admits (p. 35) that "the prejudice of the lower classes against vaccination, on account of the supposed communicability of leprosy by it, was, until the recent epidemic of small-pox in Trinidad, insurmountable."

UNITED STATES.

The origin of various recent outbreaks of leprosy in the United States is veiled in obscurity, only because medical men do not know where to look for it; though they have admitted that, since the discovery of the *bacillus lepræ*, vaccination appears to be the most probable cause. Reluctant to impugn the Jennerian practice, and with opportunities at command, they have preferred to extend their researches in other directions.

Dr. H. S. Orme, President, State Board of Health, California, considers the ordinary explanations given to account for the dissemination of this disease inadequate, and observes:—" Not heredity, nor syphilis, nor endemic conditions, could have given rise to the group of sixty cases in the village of Spain; to the outbreaks in New Brunswick and Cape Breton Island; to the sixteen cases at Charlestown between 1846 and 1876; to the forty-two now at New Orleans, or to the two at Galveston. It is often impossible to trace the source and mode of contagion, but the same is true with all the disorders whose contagiousness is disputed."—*Leprosy: its Extent and Control, p. 29*.

In the "Transactions of the Medical Society of the State of California" for 1890, vol. XX., Dr. Orme, referring to the general vaccination of the people of the Sandwich Islands in 1868 with human lymph, consequent upon an outbreak of small-pox, says :—" This reckless practice doubtless contributed greatly to the spread of both syphilis and leprosy."

"At a meeting of the New York Academy of Medicine, June 6, 1889, Dr. Prince A. Morrow gave an account of his personal observations on Leprosy. Referring to his visit to the Leper Asylum at Molokai, he said he considered it probable that a number of the cases which had arisen in the Sandwich Islands had been caused by impure vaccine."—*British Medical Journal, Aug. 31, 1889.*

Dr. T. B. Sutliff, of Sacramento, California, has sent me the following narrative of a case of invaccinated leprosy, contracted in California :—

" Boy, native of California ; never been out of this State ; no family history of any constitutional disease ; was well until about eight years ago ; was then vaccinated. The arm became very sore, and was swollen in its entire length. An abcess formed in the axilla, and subsequently broke, discharging pus freely. Recovery ensued, the arm becoming well again. Soon after it was noticed that there were numerous patches of 'ringworm' on the body. This condition continued for several months. Several persons, himself included, were vaccinated by the father directly from the boy. The father, and one other of the persons vaccinated, subsequently had 'ringworm.' The father has since had 'tetter' from time to time. One year after vaccination the boy's ears became slightly enlarged. The nose was noticed to have become broadened, and tubercles formed on it and on other parts of the face.

" The ears are markedly enlarged, the nose broadened, and the *alæ* are thickened. The hands show the disease quite plainly, the fingers being clubbed. There are a few small ulcers. There is a sore on the shoulder. The feet are beginning to be affected."

In the report of Dr. H. W. Blanc, Professor of Dermatology and the Chief Sanitary Inspector for the city of New Orleans, November 27, 1889, addressed to the President of the Board of Health for the State of Louisiana, where leprosy has been provokingly prevalent, it is stated that leprosy, syphilis, and tuberculosis are transmitted by vaccination. Dr. Blanc says that, "in his two-fold capacity of Dermatologist to the Charity Hospital of the city and Chief Sanitary Inspector of the city, he has had unusual opportunities for the study of leprosy and vaccination. During the past eight years he had observed over sixty cases of *bona-fide* leprosy (anæsthetic and tubercular). The disease is slowly increasing, it is inoculable and communicable by vaccination, and humanised virus should be avoided. Most of the vaccinations are performed by me and my assistants, and I will not on any account allow humanised virus to be used."

The Occidental Medical Times, Sacramento, California, of September, 1890, publishes "An Interesting Case of Anæsthetic Leprosy apparently following Vaccination," by Sidney Bourne Swift, Resident Physician, Leper Settlement, Molokai, H.I., and D. W. Montgomery, Professor of Pathology and Clinician for Diseases of the Skin in the Medical Department of the University of California. The writers say that "one of the most interesting points in this case is that Peke (the leper) had been vaccinated one year before developing symptoms of leprosy, and that the vaccination scar became anæsthetic. Might it not be that with the vaccine virus the virus of leprosy has also been inoculated?"

This question is answered by Dr. Chr. Uronwald, Chairman of the Sanitary Committee on Leprosy, Minnesota, Wisconsin, U.S., who says in the official report of the State Board of Health, "Vaccination has undoubtedly originated leprosy."

In a paper read before the California State Medical Society, in 1881, Dr. A. W. Saxe gives an instance of three children in Honolulu, born of American parents, who became lepers.

In consequence of the serious development of leprosy in Hawaii, there has arisen, during the past two or three years, a determined opposition to vaccination, to which the increase of leprosy is naturally attributed. The President of the Board of Health, the Hon. David Dayton, in the Report to the Board of Health, Honolulu, for 1892 (p. 27), says "that notices were inserted in the newspapers offering vaccination, but there were no applicants." In the same report (p. 67) Dr. R. B. Williams, Government Physician for Hilo and Puna, Hawaii, says:—" There is among natives great prejudice and opposition to vaccination. . . . Hence voluntary vaccination is almost impossible." The existing law renders vaccination obligatory upon all the inhabitants of these islands, and has heretofore been enforced with great rigour, particularly during outbreaks of small-pox. During the past two years it appears to have been relaxed, and objectors have been humanely suffered to escape the dreaded ordeal. It is interesting and instructive to note the result. On page 41, the President of the Board of Health, observes:—" On December 31st, 1890, there were 1213 lepers in the custody of the Board, that being the highest number ever reached; and on March 31st, 1892, there were only 1115, a decrease of 98 during the period." A recommendation is thrown out in the report, that the vaccination law should be amended, as the present laws are entirely unsuited to the times. In view of the evidence brought before the Royal Commission on Vaccination in London, showing the futility of vaccination as a preventive of small-pox, and its fertility in disseminating every inoculable disease, the only reasonable alteration in the vaccination laws that will be acceptable to an enlightened public opinion is their entire and permanent abrogation.

Personally I have heard of many cases, which I have no doubt have been due to leprous vaccination. There is, however, it is to be regretted, manifest reluctance on the part of the profession to look for causation in this direction; it is generally pooh-poohed as soon as mentioned; and when it appears to be the only way of accounting for a particular case, the attending physician prefers to hold his peace rather than discredit a practice which he has been educated to believe is the greatest discovery in the history of medicine. Professor Gairdner, of Glasgow, introduces his cases of invaccinated leprosy in the *British Medical Journal* with an apology and evident reluctance.

It is generally admitted that, given the right conditions as to environment, temperament, or idiosyncrasy, all bacterial diseases are transmissible from one human being to another; and as no single authority that I have met with, since the discovery of Hansen, pretends that leprosy does not belong to this category, the danger of vaccination is obvious. I am aware that this objection is attempted to be met by the introduction of animal lymph; but animal lymph is admitted to be too active, especially in tropical countries, to be used direct; and in general, therefore, it is available only after one or two removes, when it carries with it diseases both animal and human, as has been shown in evidence before the Royal Commission on Vaccination.

BRITISH INDIA.

An uneasy feeling is beginning to be exhibited in India on this momentous subject, owing to the accumulation of evidence tending to show the sinister connection between the extension of the State-provided remedy against small-pox and leprosy. The synchronicity between the spread of leprosy and the extension of vaccination has given rise, in some districts, to such a dread of vaccination, that every device is resorted to by thoughtful parents to prevent their children being vaccinated. Attempts have been made to remove the dread of leprous inoculation by the substitution of cow, calf, sheep, lamb, and donkey lymph; various compounds (one described by the medical purveyor as the Madras paste, and another as Lanoline) have been introduced ; and some of the leading journals now energetically demand a safer and better system of vaccination. Dr. S. N. Boral, Chief of the Vaccine Department in the Jubbulpore district, has come to the rescue of the Jennerian *cultus* in the columns of the Allahabad *Morning Post*, but he sees clearly the weight of the incriminating testimony, and admits that to deny the possibility of vaccinal syphilis or vaccinal leprosy would be tantamount to denying the value of human testimony altogether.

The authorities in India are well aware of the widespread repugnance to vaccination in that country, and of the cause of this repugnance, in the mischievous results known to every vaccinator ; but all mention of these evils is carefully omitted in their official reports. Now, however, that the dangers attending the most carefully conducted vaccination have been so fully disclosed in the voluminous evidence taken before the Royal Commission on Vaccination, the culpability of such reticence is inexcusable. *The Statesman*, of Calcutta, August 22, 1891, commenting upon the "Resolution on the Statistical Returns of Vaccination in Bengal for 1890-1," in the *Calcutta Gazette*, August 21, 1891, observes:—" It has been stated that one of the greatest objections to vaccination among the natives of India, and other Oriental peoples, is that diseases such as leprosy, and other terrible blood diseases, have been inoculated with the vaccine virus. We think that some opinion should have been expressed under this head, and the omission of it is to be regretted in what is supposed to be a report given for the general edification of the Government as well as the public." And in a leading article, November 22, 1891, this same influential Indian journal, referring to leprous vaccination, observes:— "There seems to be no possible room for doubting the reality of the very grave danger to which attention is drawn. . . . It is notorious that inoculation, that is, the direct introduction of the virus into the blood, is the chief, if not the sole, means by which leprosy is communicated. Throughout the greater part of Europe, at least, and in all the principal British Colonies and Dependencies, including India, vaccination is not merely the most common means of inoculation, but in most of the countries in question it is a means which, practically, is universally adopted, and enforced by legal penalties. It seems, then, to follow, almost of necessity, that, unless special precautions are taken to prevent so terrible a calamity, leprosy, wherever it prevails in these countries, must inevitably, in a certain proportion of cases, be communicated through the medium of vaccination. . . . When vaccination from the human subject is practised, it is quite possible that, in a few generations, it might lead to an enormous multiplication of cases, and that without implying any want of ordinary care on the part of operators. For it should be remembered—and herein, it seems to us, lies the real gravity of the danger—the disease is not one that commonly shows itself in infancy. The child from whom the lymph was taken might, to all appearance, be perfectly healthy, and yet its blood might be infected with this fatal and loathsome poison, and the operation thus make it a focus of contagion."

The Calcutta Daily News, August 7th, 1891, referring to the allegations that vaccination is responsible for the transmission of insanity, leprosy, and other diseases, observes :—" The question is an important one ; and while it is a great gain to humanity to even modify such a scourge as small-pox, it is largely discounted by the consideration that immunity from it may be purchased at the price of other diseases as bad or worse than that affliction itself."

The Bombay Guardian of April 6th, 1889, commenting upon the spread of leprosy by vaccination, observes :— " If we have to choose between the danger of leprosy and small-pox, let us by all means have the latter. The ghastly sights to be seen in every Indian public thoroughfare, of the scabious, handless arms, and footless legs of begging lepers, forbid any other alternative.

Small-pox is bad, but leprosy is a hundred times worse."[1]

When similar charges have been made against vaccination in Europe, the usual course of the official propagandists is to ignore the terrible indictment as long as possible, and then, when questioned in Parliament, either to minimise its character and attribute the results to other causes, or to deny them altogether. Those who are concerned for the public well-being in India may study with profit the pages of the Third Report of the Royal Commission, to see what lengths the official supporters of vaccination are prepared to go in their advocacy of the Jennerian *cultus*.[2]

---

[1] The *Indian Spectator*, Bombay, December 27, 1891, in a leading article, calls attention to a strong reaction that has set in against vaccination, and points out that the opponents of the Jennerian method of preventing small-pox contend that vaccination is answerable for much of the spread of leprosy in recent times, and that this view, which has the support of men considered as high authorities on the subject, is of vital importance. The article concludes by observing that "The Leprosy Commission seems to have thrown away a fine opportunity in omitting to direct its researches into the alleged connection between vaccination and leprosy."

[2] On December 30, 1880, fifty-eight young recruits belonging to the 4th Regiment of Zouaves were vaccinated from a Spanish child at the Dey Hospital, Algiers. In a few weeks all the vaccinated developed syphilis, to which about one half subsequently succumbed. The Minister of War, General Farre, instituted an inquiry, but no report has been published. Five times questions were submitted in Parliament with a view of eliciting the facts. The President of the Local Government Board promised to institute inquiries through the Foreign Office, but, on receipt of a reply, excused himself from giving particulars by declaring that "the information was incomplete," and promised to make another application. Finally, on the 27th October, 1882, Mr. C. H. Hopwood repeated his question, which was answered by Mr. Dodson (Lord Monk Bretton) by a categorical denial of the facts. At this time *the Local Government Board were in*

A report of a "Scheme for obtaining a better knowledge of the Endemic Skin Diseases of India" has been prepared by Mr. Tilbury Fox, M.D., F.R.C.P., Fellow of University College, Physician of the Department for Diseases of the Skin at University College Hospital, London, etc., and Dr. F. Farquhar, Surgeon - Major, Bengal Medical Service. Under the title of Propagation these authors include inoculation and vaccination, and observe that there is by no means a slight body of facts which seem to show that the inoculation with matter from a leprous sore—and this may occur in cohabitation and constant contact, and in vaccinations—may give rise to the disease. The authors propound a series of fifteen questions, with the view of elucidating the presence and cause of leprosy in different districts and individuals. It is noticeable that, while they allow that vaccination is a cause of the propagation of leprosy, inquiry on this point is not demanded. The Secretary of the Leprosy Investigation Committee, in his address before the International Congress on Hygiene and Demography, states that the cases of invaccinated leprosy are few in number. But, inasmuch as all the cases have cropped up accidentally and not as the result of research, it is impossible to estimate their number. An incalculable service to humanity will be performed when medical practitioners of high qualifications, not committed to any preconceived theory, will,

---

possession of full details of this vaccine tragedy (supplied by a Member of Parliament), including the names of the unfortunate sufferers, their grade, matriculation numbers, and nationality, as furnished by a medical witness, after a personal and painstaking investigation of the facts at the Hôpital du Dey.

in the interest of the public health, undertake this important investigation.

The British Medical Journal, Sept. 19th, 1891, reports an address delivered before the British Association, on " Leprosy and Vaccination," by Dr. R. Pringle, surgeon-major, late of the Sanitary Department of the Bengal Army, who said " that of all charges which had been brought against vaccination, none approached in seriousness — especially as it related to India—the charge that the present admitted increase of leprosy in certain countries was due to the increase of vaccination, which was stated to be not infrequently in those countries little else than leprous inoculation." Dr. Pringle, in recording this statement, fully admitted that in the main it rested on the evidence of medical officers of the highest rank and authority in the West Indies, and that one Indian medical officer (Surgeon-General C. R. Francis) distinctly admitted, in an important public document, that this was not only possible but probable. Surgeon-General Francis, in the *Journal of the Leprosy Investigation Committee*, No. 1, August, 1890, had written " that leprosy may be propagated by inter-marriage and hereditary transmission is undoubted. I believe, too, in its propagation by vaccination or inoculation, but I am very sceptical as to mere contact being the cause. . . . I would advocate an investigation into the effects of vaccination, there being some who are still dubious on this point, though the statement made two or three years ago by Professor Gairdner on the subject would seem to be conclusive in favour of vaccination as a factor." The inoculation alluded to by Dr. Francis was stated by Dr. Pringle not to be inoculation by flies, as seen in ophthalmia—a fact which he remarked in passing was both recognised and dreaded by the Jews of old as one of the most certain and probable means of spreading leprosy—contact, as pointed out, being to them a Levitical uncleanness, and not, therefore, necessarily risking the consequences of contagion. The inoculation alluded to by Surgeon-General Francis was small-pox inoculation, as practised in the Himalayas as a preventive against spontaneous small-pox. Dr. Pringle severely criticised the circumstances attending vaccination in the case recorded by Professor Gairdner in the *British Medical Journal*, June 11th, 1887. It was hardly to be expected that the opponents of vaccination would fail to take advantage to the full of such damaging statements regarding vaccination in India as recorded by Surgeon-General Francis; but Dr. Pringle stated that Surgeon-General Francis' experience was entirely the reverse of his own. The main object of his paper was to point out the means available for subjecting these two conflicting experiences to the most searching investigation on the spot by the Leprosy Commission now in India."

Dr. Pringle proceeds to explain that for twenty years he had collected vaccine lymph in Terri, in the Himalayas, where leprosy was very gravely endemic, and used this lymph in districts where the population was 500 to the square mile. Over two millions of these vaccinations had been performed, and he had never seen a case of leprosy traceable to vaccination. Dr. Pringle omits to say whether he ever searched for such cases, or inspected the vaccinated subjects a year or two after the operation, or inquired whether those attacked with leprosy traced their affliction to vaccination. As Dr. Pringle urges upon the Leprosy Investigation Commission the importance of inquiry to test the validity of the charges now brought from all quarters, it would seem that he had not made this investigation himself.

Dr. Vandyke Carter, of Bombay, allowed to be one of the greatest authorities on the subject, includes vaccination among the list of causations, of which he says he has found recorded a few affirmative examples, at least, of each one method or means.—*Report on Leprosy and Leper Asylums in Norway, with reference to India, p. 178.*

Mr. H. Brown, of Simla, writes to me Oct. 2, 1889:—
" Experiments have proved leprosy to be inoculable. There must necessarily be a dread of vaccination in India, since the subject from whom lymph is taken for the operation may be a leper. In India native village vaccinators are not over careful from whom they procure their lymph. In Malabar and various districts of the Madras Presidency, where I have lived hitherto, it has been no uncommon thing to see the lymph extracted from the arms of itchy native boys. Leprosy has increased alarmingly since the introduction of vaccination. I know of some cases where perfectly healthy persons, whose parents also are healthy, and who have been accustomed to live in healthy localities, have been smitten with leprosy after vaccination. I can also quote cases similar where children have died, or have become very seriously ill, immediately after vaccination, from hideous eruptions and swellings. One lady, the wife of a respectable merchant in Cochin,

assured me that *vaccination* and *nothing else* killed her baby. It was perfectly healthy until vaccination. The Surgeon-General, Dr. Brodie, in his report for last year on the 'Distribution of Diseases in the Presidency,' declared that syphilis was on the increase throughout. This increase is coincident with the introduction of compulsory vaccination in a large number of municipalities, and with the more energetic action on the part of vaccination officers in the Madras Presidency."

Dr. Chunder Ghose, in medical charge of the Leper Asylum, Calcutta, in a communication to the Secretary of the District Charitable Society of Calcutta, dated August 21st, 1889, states his opinion that leprosy is increasing in India, that it is communicable by vaccination, and that there is a dread of the operation on this account.

Dr. Roger S. Chew, of Calcutta, who was for six years in the Medical Department of Her Majesty's Army in British India, and has devoted fourteen years to the study of leprosy in India and other countries, furnishes, in his pamphlet on leprosy, a table giving the results of his investigations into the causation of the 1034 cases which have come under his treatment. Of these, he says, insanitation is responsible for 105 cases, vaccination for 148, and 72 cases are due to other forms of inoculation. Dr. Chew has been kind enough to supply me with the following particulars from his case-book, in which, according to his diagnosis and careful inquiries, the disease is directly traceable to vaccination.

The cases here quoted are samples of a large number collected by Dr. Roger S. Chew, in which the connection between the onset of leprosy and a previous vaccination is well determined. With reference to this, I am advised by high medical authority that a secondary development of the results of an inoculation frequently coincides with a re-awakening of disturbance at the point of inoculation. *Also, that when a secondary and constitutional disease first indicates its existence at the scar of vaccination, it may be taken as conclusive evidence that it is consequential to that vaccination.* It will be noticed that, in almost all of these cases, the place of inoculation is first attacked, and in all of them it is affected.

Another point in the evidence adduced should be noted, namely, that it does not seem to matter whether the vaccination "took" or not for the secondary effects to be manifested in due course. The probability is that, if a vaccination is immediately "successful," a portion of the poison is discharged, and the rest retained. If there is no immediate result, the poison may still lie dormant for a variable period, concerning which practically nothing is known.

Extracts from Memoranda in case book by Dr. Roger S. Chew, Calcutta, showing connection between vaccination and the commencement of leprosy:—

1.—"Jahoorie, aged twenty-eight. Married; no children. Duration of leprosy, twenty years.

"There is no history of syphilis, either with himself or his relatives. When he was about seven years old, he was vaccinated on his right arm. About six months after he noticed a white patch over vaccine site; a similar patch appeared on his right buttock, and he soon after lost sensation in his left foot. The marks gradually faded away, broke out afresh in other portions of his body, and again disappeared to reappear, *et seq.*; but wherever these marks appeared, they were accompanied by loss of sensation, which remained permanent throughout. About sixteen years ago he suffered from enlarged spleen, for which he was fired (*i.e.*, burned with hot iron). Ten years ago the fingers of both hands began to be flexed on themselves.

" Present state :—Perpendicular of ankylosed fingers is $\frac{1}{4}$-inch on right hand, and $\frac{7}{8}$-inch on left. Fingers and thumbs of both hands much ankylosed, discoloured, and anæsthetic. Dry ulcer at inner flexion of right thumb. Open and entirely painless ulcer on tip of nose, extending $1\frac{1}{4}$-inch inwards. Anæsthetic discoloured patches all over chest and abdomen. Two large (burn) scars over splenic region. Right foot has three large open sinuses, freely discharging a viscous, stinking, purulent fluid. On inserting little finger into the largest of these three sinuses, free vent was obtained, and the finger, striking against the astragalus, not only caused the patient a great deal of pain, but also brought away a quantity of fœtid caseous-looking matter, due to caries of the bone. There are three discharging sinuses — the largest one inch in diameter at the opening—in the left foot also. Raw, ulcerated, and entirely painless stump, marking where middle toe of left foot has recently fallen off. Second toe of same foot has also dropped off, and on outer side of little toe of same foot is a peculiar blistered surface. Both shins largely covered with scaly and desquamating cuticle, with absolute loss of sensation. Hacking cough, and great pain in left thorax, with frequent and bloody *sputa*, beginning some two years ago, but aggravated during the last month. Caseous degeneration at extreme apex of right lobe of left lung.

2.—" Daidas, a male, aged forty-three ; a native palki bearer. Duration of leprosy, twenty years.

"There is no history whatever of syphilis or syphilitic or leprous hereditation, nor are any of his relatives similarly afflicted. He was forcibly vaccinated about twenty-one years ago. The operation was not successful, but a year afterwards the vaccine pock grew rough, lost sensation, and gave place to a small crop of *papillæ*, which spread, grew larger, and became entirely anæsthetic. He never had fever of spleen. About four years ago he used to get peculiar cramping sensations all along the course of the right ulnar (nerve) accompanied by extremes of cold and heat in fourth and little fingers. Contractions began to supervene, and these two fingers, entirely losing sensation and their power of grip, became permanently flexed on themselves. At this time his brows began to protrude and his upper lip to get thick.

"Present condition :—Permanent and rigid ankylosis of fourth and little fingers of right hand, with total anæsthesia. Tumoid, discoloured, anæsthetic patches and papules all over hand and arm from finger tips to clavicular articulation. Large tumoid and anæsthetic patch, 5 × 4 in., on anterior aspect of right shoulder; another, 3 × 1 in., over vaccination site; and a third, 3 × 2½ in., on elbow of same side. Upper lips very tumoid with paralysis of left side and partial anæsthesia of right. Uvula elongated and anæmic; fauces very anæmic. Sense of smell and taste are unimpaired. Left inguinal glands are indurated and slightly enlarged. Scrotum and prepuce elephantoid and anæsthetic. Posterior portion of scrotum is ulcerated. There are a few white spots over buttocks, whitish mark over left second toe, with loss of feeling. Ears are rather swollen. Features leonine, the supraorbital ridges being very tumoid and perfectly anæsthetic.

3.—"I. Chundar Ghoral, a male, aged forty-two; a Brahmin, and farmer by occupation. Duration of leprosy, twelve years.

"There is no history whatever of heredity, syphilis, impure living, or irregular habits. He was vaccinated seventeen years ago (the operation did not take), and five years afterwards a small rash broke out on that portion of his right arm where he had been punctured. This rash spread, ulcerated, healed, and left behind a thick warty lump, 3½ in. long and 1¾ in. broad, which was utterly devoid of feeling. Similar excrescences appeared on various parts of his body, and about seven years ago he noticed some peculiar white patches appear.

"Present condition:—Open sinuses under ball of both great toes. Nodes all along course of lymphatics. Anæsthetic patches over several parts of body. Excoriation on left elbow. Leucoderma in patches of rather large areas in different parts of back and thorax. Features leonine and feet slightly swollen and œdematous. Knees swollen and marked with tubercular ridges. Complete loss of power and sensation n the fingers of both hands, but no ankylosis.

4.—"Rohim Bux, a male Mahomedan, aged twenty-five, a hackney carriage driver, stated on the 18th August, 1889, that, a fortnight after vaccination, which did not 'take,' eighteen years ago, his parents noticed whitish patches occupying the site where the vaccine pits should have appeared, and remaining persistent for two years, after which they increased in size, coalesced, and steadily kept increasing to their present extent. . . . On his forehead, half of right arm and hand, and on his right shin, are pinkish-white patches that do not itch; leprous ulcers on his right foot, implicating the great toe, the nail of which has sloughed off. A painless ulcer, $2\frac{1}{2}$ inches by $1\frac{3}{4}$ inches, occupies the surface of right hip joint. These ulcers are all perfectly anæsthetic. The ulcer on his hip joint is nearly three years old, scabbing, and breaking out afresh at irregular intervals.

5.—"Meer Mahomed, a Mahomedan male, aged thirty-four years, married, with eight children; a clerk by profession. His parents were quite healthy, and so are his wife and six of his children, but two daughters are lepers. On the 23rd August he showed bright silvery lines on the palmar surfaces of both hands, while on the dorsal surfaces were several small annular patches, together with three large crusted patches, the largest 1 in. by $\frac{3}{4}$ in. First joint of left fourth finger had sloughed off, leaving a very angry stump. Running ulcers in right leg and thigh. Angles of mouth tough and thickened. Both ears nodular and tumoid (thickness, $1\frac{1}{4}$ in.). On chest and left arm were white patches, three of which formed a peculiar triangle which covered the vaccine pits, *in which the disease first appeared.* The lymph for the operation was taken from the arm of an apparently healthy native child. Patient has never had syphilis, nor does he know of any member of his family being afflicted with any blood disorder.

6.—"Bundaban Mullick, a Hindu male child of seven years, exhibited on the 27th August, 1889, several leprous ulcers on left wrist and at the angle of his mouth. On left arm was an oblong whitish-pink patch that entirely obliterated the vaccine pits. The boy's father, who is perfectly healthy, says that two and a half years ago the boy was vaccinated, and about four months after the operation he noticed the puncture sites occupied by three small white patches, which in the course of one year extended and coalesced to form the single patch now seen. The ulcers appeared about a year ago, and refused to heal up.

7.—"A Hindu male, Bhaleshur, aged ten years. Six months after vaccination, three and a half years ago, a white patch appeared over his left clavicle and on vaccine site. Half of the patch on clavicle ulcerated, and the ulcer, ¾-inch broad and 1½ inch long, refused to yield to either arsenic or mercury, both of which he had taken for two years. The clavicle is denuded of flesh, and plainly visible to the naked eye, while the ulcer itself is of a leprous type.

8.—"A Hindu female, Gowrah, aged nineteen, stated that she was vaccinated in her seventh year (as far as she remembers). The operation did not 'take,' and, five months after, three whitish spots appeared on her right arm, where the vaccine pits should have been. These spots extended, coalesced, and, spreading downwards, disfigured her arm as far as the wrist. Her lips thickened, menses became irregular, and obstinate sores, which are still open, broke out on her feet.

9.—"Vincent D'C., an East Indian clerk, aged thirty. A year after vaccination (six years ago) he felt a peculiar constant itching in the vaccine pits, and, a short time after, noticed on his left arm a curious rash, which subsequently gave place to obstinate ulcers, for which he was unsuccessfully treated by three different doctors. On his left arm are three irregular-margined annular patches, averaging 1½ inches in diameter.

10.—"Da Singh, a Hindu schoolboy, aged twelve, told me (Dr. Chew) on the 11th December, 1889, that he had been vaccinated in his seventh year. About eight months after the operation, the vaccine pits ulcerated, and the ulcers spread and coalesced, to form a large annular patch which has obstinately remained open. On same arm (right), a little above the elbow joint, appeared a long pinkish white patch, which gradually enlarged, and now occupies a space, 9½ × 3¼ inches wide, running over flexor aspect of portions of both arm and forearm.

11.—"Francis G——, an East Indian male child, aged three ; admitted on 21st December, 1889, has eczema of the scalp of three months' duration. He was vaccinated when only eleven months old. The operation did not 'take,' and a month afterwards the whole of that left arm became perfectly anæsthetic. An annular patch of the size of a rupee occupies the left angle of his mouth, and close by this annular patch are two small anæsthetic tumoid ridges.

12.—"Ameer Hoosein, a Mahommedan lad, aged fifteen, was vaccinated in his seventh year. The operation 'took,' but in his eighth year the vaccine pits turned into little ulcers, which enlarged, scabbed, cleared, and finally coalesced to form one large annular patch on his left arm. Smaller annular patches exist on left shoulder, right hand, and right foot.

13.—"Doorie, a male, aged 18, is a leper, and attributes his disease, from which he has suffered for the past nine years, to impure vaccination. The vaccine pits are badly ulcerated, and he has an anæsthetic patch covering the entire elbow joint of the same arm. Both lungs are implicated.

14.—"Kasmini Bibi, a married Mahommedan female, the mother of two healthy children, deposed, on the 29th December, 1889, that she was vaccinated on her right arm in her thirteenth year. The operation 'took,' but five years afterwards the pits ulcerated, and these ulcers remained obstinately open. At this time crops of *papillæ* accompanied by total anæsthesia, began to show themselves on her chin, right breast, right thigh, and right knee. About one and a half years ago a tiny white spot appeared in the centre of her forehead, and this spot has now increased to a circular patch the size of a shilling piece.

15.—"Goolburee, a married Mahommedan female, aged forty-seven, with pemphigus (five blotches on right foot) of one and a half years' and leprosy of thirteen years' standing. The latter disease she attributes to improper vaccination. She has been five times vaccinated—the last occasion when thirty years of age. Four years after this the leprosy manifested itself. There are nodosities in both ears, in her nose, on her left arm, and along the left sterno-cleido-mastoid; one ulcer, $1\frac{1}{2} \times 1\frac{1}{4}$ inches, on right ankle, and another twice this size covers last vaccine site on right arm.

16.—"Mabel P——, a Scotch lassie, aged seventeen, and a leper for the last eight years, was brought by her mother, who stated that she was vaccinated when she was seven and a half years old. About six months after the operation, which was successful, symptoms of leprosy began to develop, and she flew here and there to every medical practitioner that money could procure to save her child, but to no avail, as the disease kept increasing. The girl's present condition is :—Ears tumefied, and 2½ inches thick. Face marked with ridges like cooled lava (volcanic). Sores on hands and feet and angles of mouth. Eyesight impaired. . . Tonsils indurated. Uvula ulcerated. . . Ankylosis of little and fourth fingers of both hands. Ulcer in left nostril, the right being entirely blocked, and external nares flattened out. Lips protruding, thick, and hardened. Slight contraction of right knee, and anæsthesia well marked everywhere.

17.—" M. T——, a Eurasian female, aged fifteen, whose father stated that three years ago—at the time of the great scare caused by an expected epidemic of small-pox—he had his daughter vaccinated. The operation "took," but a year after the pits ulcerated and refused to yield to treatment. At the same time a few white spots appeared on her back, her sides, the nape of her neck, and over her face. Those on her face grew larger, till, impinging on each other, they finally coalesced to form one large blotch of pinky-white, which, contrasting against her olive brown complexion, terribly disfigured her.

18.—"A. A—n, a Chinese carpenter, aged forty-three, married, and the father of four healthy children, deposed on the 22nd December, 1890, that four months after vaccination, eleven years ago, the vaccine pits broke down into ulcers, which are still open. His ears are tumoid (1¼ inches) and perfectly anæsthetic. Tubercular deposit, and well marked anæsthesia in patches all over body.

19.—" Imrato J. Ghose, a Hindu male, aged twenty-eight years, stated, on 14th Feb., 1891, that he had been three times vaccinated, in infancy, at six years, and at twenty-one years of age, at each of which times the operation was very successful. Fifteen months after the last vaccination his leprosy showed itself. His body is covered with hypertrophied anæsthetic patches of various sizes and contiguous to each other. Mercurial fœtor of breath. Ears tumoid. *Alæ nasi* partially affected, and anæsthetic. Brows very slightly leonine ; pains in loins in rising, and drawing up the legs ; ulcer on ball of great toe of left foot, freely discharging serum. Vaccine pits are badly implicated in the hypertrophies on left arm.

20.—"Sibhoo, a Hindu male, aged forty, and a widower, stated on the 14th March, 1891, that he was vaccinated when thirty-six years old. A year afterwards the vaccine pits ulcerated (ulcers still open). He has never had syphilis in any stage; his people are healthy; his body is well nourished, with patches of discoloured (whitish) hypertrophied and anæsthetic cuticle—the smallest 4 × 5 inches—scattered all over. Black anæsthetic patch over surface of lower half of left "tensor vaginæ femoris." Little toe of left foot has dropped off. Both ankles are much swollen and inflamed; there is entire loss of sensation in fingers, feet, and ulcers. Under the balls of both great toes are running ulcers.

"21.—William J. C——, an East Indian male, aged twenty-one, admitted on 31st August, 1891, was vaccinated three years ago, and a year afterwards noticed some pimples break out in vicinity of the pits, on back of hands, and on his back. These at first itched a great deal, but, breaking down into pustules, became devoid of feeling, and extended to various portions of his body. The left, fourth, and little fingers are also anæsthetic.

22.—"Khyroo, a Mahomedan male, aged fifty, applied for treatment on the 20th October, 1891. Has been married twenty years, and has four male and two female children, all healthy. Leprosy showed itself nearly a year after he was vaccinated. Prior to this he was always hale and hearty. Had small-pox eight years ago (*i e.*, four years after he was a leper). Present condition:—Ears are slightly tubercular, their hypertrophy measuring about $\frac{1}{32}$ inch, while their long and short diameters are normal, without sore or abrasion, and with slight anæsthesia. Face marked with small-pox pits, hypertrophy over malar ridges, angles of mouth and bridge of nose, the *alæ* of which are thickened, and tubercular anæsthesia is well marked. Arms covered with patches of tubercular deposit and vesicles, with well-marked anæsthesia.

"Hands—Thumb, ring and little finger of right hand permanently flexed; burn on thumb and forefinger, proving anæsthesia; nails deformed and splitting up; tubercular infiltration. Left forefinger much swollen, and a small ulcer on back of first joint, which is bent. Little finger also swollen and flexed on itself; fourth finger slightly flexed. Anæsthesia well marked in all the fingers, and in the palms and wrists. Small patches of tubercular growth all over wrists and dorsal surface of hands.

"Body, legs, and feet all show well-marked developments of tubercular leprosy, with anæsthesia.

23.—"G. D'R——, an East Indian of thirty years, admitted 3rd January, 1892, stated that one year after vaccination his disease appeared. Present condition:—Ears wrinkled and elongated, with irregular nodes. Features leonine. Brows tumoid and tubercular, malar ridges hypertrophied and anæsthetic. Tubercular deposit at angles of mouth; nose flattened out, anæsthetic, and depressed at bridge. Tubercular deposit and anæsthetic patches along the entire length of both arms. Fingers contracted, ulcerated, and anæsthetic. Body also affected. Right leg elephantoid; the left anæsthetic. Feet well marked with the disease, being ulcerated and mutilated.

24.—"R. B. M——, a Brahmin of thirty-eight years of age, stated that three years subsequent to vaccination the pits ulcerated and became anæsthetic. Features leonine, angles of mouth hypertrophied and anæsthetic. Arms ulcerated, with hypertrophied cuticle : ankylosis of elbow joint of left arm. Fingers of both hands badly ankylosed. A few patches of leucoderma on his back, and one on thorax. Large anæsthetic sinus under ball of great toe of both feet.

25.—"Ishar Ghosal, a male Hindu, forty years of age, admitted 3rd January, 1892, deposed that he was vaccinated when twenty years of age, and five years afterwards the pits ulcerated. The ulcers healed, their site entirely lost feeling, and the anæsthesia spread till it implicated the entire length of right arm. Present condition:—Features slightly leonine; face shows scars of old sores; hypertrophy and anæsthesia of malar ridges and angles of mouth. Besides the anæsthesia of right arm, there is a large whitish patch (5 × 2¼ inches), just above elbow. Ankylosis of fourth and little finger of both hands. Large patches of leucoderma on back, and one large white patch covers the entire thorax. The interspaces are hypertrophied and anæsthetic. Feet badly ulcerated, the ulcers being anæsthetic and discharging freely."

Dr. A. Mitra, Chief Medical Officer, Kashmir, says:— "I have on three occasions searched for bacilli. In one instance I found them in lymph from a vaccinated leper."—*American Journal of Medical Sciences, July, 1891.*

In the year 1888 the Government of India adopted and issued a series of resolutions on the subject of leprosy, admitting the increase of the disease, and acknowledging the impossibility of dealing with it in any effective manner. No word of warning was then, or has been since, uttered as to the inoculability of the disease, now so generally admitted by medical authorities, and the consequent dangers of its dissemination at the hand of the public vaccinator, nor is it likely that this source of infection will be officially condemned until the people in India, as in many parts of England, stand upon their parental rights, and refuse at all costs to imperil the health and lives of their offspring by this irrational and disease-engendering rite.

Some medical authorities, while admitting that leprosy is inoculable, and disseminated by vaccination, insist, for the credit of the Jennerian practice, that such cases are very rare, and are due entirely to the carelessness of the operator, and that, therefore, it is unreasonable to throw discredit on so beneficent a discovery on this account. This mode of reasoning may satisfy the unreflecting; but if it be once allowed that leprosy is transmissible by vaccination, who can estimate the extent of the resulting mischief? Vaccination is practised in all the colonies and dependencies of our empire, and in all countries where leprosy prevails. The disease is usually of slow incubation, and, until external indications of the malady are exhibited, a child may be, and often is, used as a vaccinifer without inquiry. Here, then, in leprous countries, are all the conditions necessary for inoculating the germs of leprosy into the blood of present and future generations. The late Dr. George Hoggan, of Beaulieu, France, who devoted many years to the study of leprosy, has examined many lepers in Europe, and he attributes the disease in nearly all cases to vaccination.

The following letter explains Dr. Hoggan's view on this part of the subject :—

Beaulieu, Alps Maritimes,
December 29th, 1889.

DEAR MR. TEBB,— Upon the connection between vaccination and leprosy I hold a very strong opinion. Apart from the opportunities which I have had in Egypt, Palestine, and elsewhere of studying leprosy in the mass, I think that my extensive researches into the minute pathology of the disease, as evidenced by the papers published in the "Pathological Transactions" for 1879 and "Archives de Physiologie" for 1882, warrant me fully in expressing a firm conviction on the subject. At pages 88 and 90 of the latter work I refer to the relations between vaccination and leprous infection, only, however, to show the difficulty of connecting the two in the history of the case. Taking all the factors into consideration, I hold that, in the cases of leprosy I was then investigating, the disease was conveyed through vaccination. I further believe that, in the majority of cases of leprosy developing in children, the leprous infection is transmitted along with the vaccine virus. In adults, on the contrary, I have had evidence that leprosy is often conveyed along with syphilis ; and this, taken in connection with vaccinal infection in the young, had led me to suggest the following explanation of infection in leprosy :—Hitherto all untainted evidence has shown that leprosy cannot be inoculated *per se* into a healthy body.[1] Combined, however, with the virus of small-pox, syphilis,

---

[1] Dr. Hoggan, no doubt, refers to the negative results mentioned by Leloir in "De la Lèpre," p. 237, where a doctor inoculated twenty healthy persons with leprous pus, blood, and tubercle, and to Profeta's

or other diseases, it seems to be easily transmissible into the system, and it is in this direction that future investigations should be pursued.—I am, dear sir, yours faithfully,

GEORGE HOGGAN, M.B.

In page 74 of the "Report of the Royal College of Physicians on Leprosy," dated 1867, is the following important suggestion:—

"The question alluded to in the communications from Dr. Erasmus Wilson and Sir R. Martin (*vide* Appendix to Report) as to the transmission of leprous disease by vaccination and wet-nursing, is one of special interest to Europeans resident in India and other tropical countries, and calls for searching examination."

The cases referred to are—Case 1, p. 235—*Elephantiasis tuberculosa*; duration of latent period, two years; total duration, five years; no pains; febrile attack, simulating rubeola; vaccinated from a native child:—

"A young gentleman, age 16, with fair hair and complexion, and somewhat more youthful in appearance than might be expected of his age, has been afflicted with the tubercular form of leprosy about five years. He was born in Ceylon, is the son of European parents, and one of six children, all of whom are healthy. His father and mother have always enjoyed good health,

---

inoculations (p. 238) of twelve persons, including himself, with blood and pus from leprous ulcers in wounds made by scarification and surfaces laid bare by blisters, and to his subcutaneous injections of matter from leprous tubercle. He overlooks, however, the numerous recorded cases of accidental leprous inoculations. Failures in intentional inoculations in countries free from leprosy cannot be set against cases of inoculation through abrasions of the skin or when the leprous poison is introduced into the blood by means of the vaccinator's lancet.

the father having resided in Ceylon for twenty years, the mother since her marriage. He was nursed by his mother, but vaccinated from a native child."—*College of Physicians' Report.*

Page 239, Case 9—*Elephantiasis anæsthetica* following vaccination (given in Dr. Erasmus Wilson's work, 1867, pp. 620-2) :—

"A lady, aged 26, the wife of an officer in the Indian army, became affected with elephantiasis in 1861. She was born in Calcutta of European parents, and brought to England when two years old. She returned to India in 1853; was married in 1855; has been eight years married, and has now (1863) revisited England for medical treatment. In 1861, then being in Oudh, she was vaccinated from a native child, and shortly after vaccination 'a slight spot came on her cheek, and increased in size to the diameter of a shilling.' It was hard to the touch, a little raised above the level of the surrounding skin, and of a dull red colour, without pain or tenderness. The swelling was painted with iodine, and afterwards blistered several times, and the blister kept open; but although somewhat reduced in size, the prominence was not removed. About six months later dull red flat spots appeared, dispersed over the greater part of her body. Her hands and feet became swollen, and she had pains of some severity in her joints and feet."

The same author gives the following on p. 650 (p. 86 of the Leprosy Committee Report) :—

"Dr. Bolton, of Mauritius, mentions the case of a boy of fourteen, afflicted with leprosy from the age of seven, the son of British parents, whose father ascribed the origin of the disease to vaccination. . . . Several medical men, who have had the opportunity of watching the disease closely, expressed their belief that leprosy may be conveyed to sound persons through the medium of the discharges of ulcers."

Referring to the series of papers which he reports, Dr. Erasmus Wilson says ("Royal College of Physicians' Report," 1867, p. 234):—"Our cases also favour the suppositions of the existence of other modes of transmission (transmission by generation—*supra*), namely, by lactation, by vaccine inoculation, and by syphilitic inoculation. The first of these methods of contagion lies beyond the reach of remedy; the others are preventible."

In the December number of the *Nineteenth Century* for 1889, p. 929, Sir Morell Mackenzie, in an article entitled "The Dreadful Revival of Leprosy," says:— "There is, or was quite lately, a boy in a large public school, in whom there are the strongest grounds for suspecting the existence of leprosy in the early stage; the disease is supposed to have been communicated by vaccination in the West Indies. It is beyond question also that there are many other cases in this country at the present moment, which are carefully concealed from the knowledge of every one but the medical adviser. Nearly every skin specialist must be able to attest this fact."

Dr. Suzor, of Mauritius, stated in the *Progrès Médical*, No. 14, that "in one instance two children of healthy parents became lepers, apparently as the result of having been vaccinated with lymph taken from a child belonging to a leprous family," and he thought the cases furnished conclusive proof of the communicability of the disease by vaccination.

In the *British Medical Journal* for Nov. 29, 1890, there is given a tolerably full report of the discussion on "Vaccination Eruptions" at the annual meeting of the British Medical Association held in Birmingham.

The discussion was opened by Mr. Malcolm Morris, M.R.C.S. (Edin.), who presented a formidable classification of vaccinal eruptions, including constitutional diseases of the most serious character, amongst which he included leprosy.

Dr. A. M. Brown observes in his pamphlet on "Leprosy in its Contagio-Syphilitic and Vaccinal Aspects," 1888, pp. 12, 16, 17 : "When we come to note a recent medical disclosure bearing on this point (inoculation of leprosy by vaccine virus) and which has been allowed to pass unheeded for the reason, I presume, that it tells against the Jennerian and Pasteurian theory and practice, the necessity for strictest caution will be obvious. Hypothesis and specific *bacilli* apart, the observations of Arning, and alas, of too many who do not care to confess it, prove that vaccination is capable of actually transmitting *lepra* from the leprous to the non-leprous. The fact is unmistakable, and our duty is to make mankind and the medical profession clearly comprehend what this implies. . . . The unanimity and persistency with which vaccination in markedly leprous countries is charged with propagating and disseminating the malady, the well confirmed coincidence of leprous centres with vaccination centres, and the discovery of the specific *bacilli* in those leprously vaccinated, ought to satisfy all who are capable of weighing evidence, or of rational reflection, that controversy on the question must, and will, ere long, be silenced."

In a communication which I received from Dr. Brown, June, 1891, he refers to the communicability of leprosy by vaccination, and the attitude of the medical profession as a body towards the ever-increasing weight of evidence. "The fact of leprosy being communicable by inoculation is clearly shown by evidence to be beyond cavil or question. Medical men, seriously interested in their profession, who fail to see this are unfortunate, and their position must be charitably attributed to indifference to the whole question of leprosy, or a dread of the overthrow of some pathological doctrine to which they are practically pledged. The evidence of the fact that leprosy is communicable through vaccination is rapidly accumulating, and the force of its importance on the medical mind, as exhibited in its special journals, is anything but welcome. The fact of the exclusion of numerous well-known cases of invaccinated leprosy from the published papers of the Leprosy Investigation Committee, though most regrettable, is not surprising; the still ignoring of experimental data by inoculating bacillo grafters is quite as little so. It is certainly amazing to find that probably the chief factor in the dissemination of leprosy in the present day—Jennerian vaccination—should have been practically set aside by the Indian Commissioners. Considering the amount of conclusive evidence now before us, many, like myself, must have felt appalled to find that this was so. Still, we may feel perfectly assured that such a performance can no more score than the proverbial Hamlet-play where the leading role has been left out."

The *Hospital Gazette*, London, Oct. 1, 1890, contains the following, under the heading, "Leprosy and Vaccination ":—"We know now that it is possible to transmit leprosy by inoculation, and it therefore behoves those who practise in countries where the disease is endemic to be very careful whence they take their vaccine. Leprosy is about as frightful a disease as any that poor man is exposed to, and beside it syphilis—the vaccinator's bugbear—sinks into comparative insignificance. A contemporary recalls the case of a native of the Sandwich Islands, who developed leprosy a year after vaccination, and seems disposed to raise the question as to whether the vaccination might not have been the means of conveying the infection."

Dr. Bechtinger, of Vienna, who has devoted thirty years to the study of leprosy in many countries, says:—"No scientific man will deny that leprosy, like all bacterial diseases, is inoculable;" and he attributes the present increase of leprosy to the vaccinator's lancet.

Dr. W. Munro says:—"I am decidedly of opinion that by careless vaccination, bloody matter being taken with the vaccine lymph, leprosy can and most certainly would be propagated. . . . I decidedly consider that leprosy can be inoculated."—*Leprosy an Imperial Danger, by Archdeacon Wright, p. 72.*

Referring to an alleged case of infection by means of vaccination, the *British Medical Journal*, October 25, 1890, says:—"Remembering Arning's important observations of leprosy bacilli in vaccine lymph taken from a leper, it is not to be denied that such inoculation may be occasionally possible." Perhaps this admission from the editor of a journal which has defended vaccination against all attacks and organised a powerful medical opposition against any modification of a most stringent and cruel law goes as far as could be expected. Mr. Ernest Hart has not thought it expedient, however, to submit the extraordinary statements in his "Truth Concerning Vaccination" to the test of cross-examination before the Royal Commission.

"Chambers' Encyclopædia," 1891, vol. 6, page 585, says:—" Evidence has recently been adduced which seems to show that it (leprosy) may be communicated by vaccination from a leprous child."

In the appendix to his work on "Leprosy," p. 274, Dr. George Thin has the following :—"In the 'Monats. f. Prakt. Derm.,' Vol. XIII., No. 1, the report of a case is extracted from the *Occidental Medical Times*, of a leper who was vaccinated in 1878, who a year afterwards became leprous, and who at present has a large anæsthetic scar at the point of vaccination. This man had healthy parents, and of two brothers and three sisters, one had died of tubercular leprosy. He is twenty-five years of age, and has mixed with lepers all his life."

In his work of 280 pages on " Leprosy " (Percival & Co., 1891), Dr. Thin devotes four pages to "Vaccination in Relation to Leprosy," and refers to Dr. Gairdner's cases. "The presumption," says Dr. Thin, ' that the disease was conveyed to the second and third child in the vaccine lymph is strong, but the case is by no means proved." Amongst others, Dr. Thin, p. 194, quotes the following from Dr. Daubler "Monats. f. Prakt. Derm.," Vol. VIII., p. 123), who 'relates two cases of leprosy at Robben Island, in South Africa, in which he believes it to be proven that the disease was conveyed by vaccination " :—

"The first case is that of a woman, H., thirty-six years old, married, and the mother of a healthy child of twelve. There was no leprosy in the family. Several years previously, on account of an epidemic of small-pox, she was re-vaccinated, the first vaccination having been effected when she was two years old. In the course of the two months following the re-vaccination, she experienced attacks of shivering and fever three to five times weekly, was frequently thirsty, but passed less urine than usual, and whilst the points of vaccination swelled and became brown, she grew dull and weak. She had been vaccinated on both arms over the insertion of the deltoid. No pustules formed, and when she saw the medical man two months after the vaccination, the parts were swollen. The swelling had begun three days after the insertion of the lymph, and reached its greatest extent eight days afterwards. At this time the parts became yellowish, and within fourteen days of the vaccination on each point there was a raised, discoloured skin, of a yellowish brown colour, and as large as a two-shilling piece. These swellings gradually increased, and, ten weeks after the vaccination, her physician found the skin of the arms and upper third of the forearm brown in colour and uneven. The brown spots extended lower down, when, after three more weeks, in which she was feverish and ill, the spots became swollen and smaller, but the skin did not resume its normal colour. In the fourteenth week after vaccination she had a violent rigor, repeated twice within the following week. Subsequent attacks of fever were at longer intervals, and not so severe. At and shortly after the severe rigors, brownish spots developed on the cheeks and forehead. Eighteen weeks after the vaccination leprous tubercles developed on the brow and on the cheeks. Two years later the woman was sent to the leper asylum at Robben Island, where she was seen and photographed by Dr. Daubler, tubercular leprosy being fully developed.

"The other case was that of a girl, fifteen years old, who was re-vaccinated at the same time and by the same medical man who vaccinated the woman H. The same local appearances followed on the arms as those described in the woman.

"After two months there were maculæ on the forehead and cheeks, and after three months more, leprous tubercles on the forehead. When seen and photographed by Dr. Daubler, the disease had lasted three and a-half years. Inquiries made at the homes of both patients, and from the medical man who vaccinated them, showed that the person from whom the lymph was taken had died of tubercular leprosy several months before, other members of the family being leprous, . . facts of which the practitioner was ignorant when he took the lymph with which he vaccinated the patients."

These cases have been carefully investigated, and their description in the "Monats. f. Prakt. Derm." is accompanied by photographs. On my visit to Robben Island, February 10th, 1892, I met Dr. P. Travers Stubbs, who is much interested in the leprosy question, and believes that the disease is inoculable and spread by vaccination. Dr. Stubbs was at that time acting as *locum tenens* for one of the medical officers at the Leper Settlement, and has since been kind enough to furnish me with the following further details concerning these remarkable cases of invaccinated leprosy :—

NOTES OF THE TWO CASES OF LEPROSY ON ROBBEN ISLAND,

AS REPORTED BY DR. DAUBLER.

Elizabeth Hart, aged 39.

European. Born at Cape Town.
Date of admission—26th April, 1887.
Race—English. Came from Wynberg (*i.e.*, eight miles from Cape Town).
Disease—Tubercular leprosy.
When contracted—When small-pox was prevalent at the Cape in 1885.
Where—
Hereditary—No.
Complications—

Particulars as taken from the Case-book, Medical Department, Robben Island.

The patient says:—" I was quite healthy until vaccinated at Wynberg by Dr. Silke in 1885. I was living with my husband in good circumstances, and had never come in contact with leprosy. About a year after vaccination a large livid patch began to appear round the vaccination mark. A few months later a creeping sensation on both sides of the face, worse on the left. Soon after this the face gradually began to swell."

Present Condition, May 3rd, 1890.

Tubercular condition of both sides of face and ears—the left more so; loss of eyebrows; some loss of hair. Tongue a little affected in front. Both hands rather swelled and tender. Infiltration of forearms and upper arms up to shoulders. Legs same as arms. Ulceration about both ankles. There is no marked anæsthesia. No special indication at seat of vaccination on left arm, which, patient says, ran its usual course.

Has been under treatment about six months. First, iodides; gurjun about four months. The appearance of patient has much improved; also general health.

February 10, 1891.

Elizabeth Hart, aged 39, married 16 years. Occupation, housewife. One son born one year after marriage. No miscarriages. Menstruation irregular. Last unwell three years ago. Her husband is still living; also the son; both at work. Her father died—the result of an accident. Mother died of phthisis, aged 40. No sisters. Three brothers—one living, healthy; two dead—one of dropsy, other found dead.

She states she was healthy until one year after vaccination, when her attention was drawn to a peculiar lividness of her left arm around the vaccination scar, which is still visible, and skin around quite healthy. There is no infiltration around the scar, which is of ordinary size and clearly visible. She consulted a doctor about this lividness, and then she went to the new Somerset Hospital, and attended for some six months, taking medicine all the time. She had pains in her legs at this time. She had no rash over her body. No sore throat, tongue, etc. Her nose, fingers, and hands became affected two years ago, 1888. Her hair began to fall out before she went to the hospital; eyelashes before she was admitted to the Robben Island Leper Asylum.

At present she has very little hair—no eyelashes and eyebrows. Her nose is very much disfigured by old ulceration. Tongue—nothing abnormal seen. No affection of her shoulder or elbow joints. Both her wrist joints are swollen, painful, and infiltrated.

There are small ulcerated spots about the size of a threepenny-piece, which are beginning to discharge; are not deep, nor are the edges undermined. The palms of both hands were sound. The last phalanges of all the fingers and thumbs were destroyed, and the finger nails are cracked and shrivelled. No anæsthesia of fingers. Her legs from her knees downwards were in a similar condition. No periostitis of tibiæ. No enlargement of glands (occipital).

In my opinion, this woman has leprosy, plus evidences of syphilis, but I am unable to find out when the latter was contracted or how.

P. B. TRAVERS STUBBS.

Ellen Wangell, 16 years.
English.
Admitted November 1, 1889, from Old Somerset Hospital.
Disease—Tuberculated leprosy.
Has lived in Cape Town and Clairmont all her life.
Contracted about 1885 at Clairmont.
Parents and brother healthy.
Tertiary rupial sores.

Patient says:—" I was quite healthy till re-vaccinated at Clairmont in 1885 by Dr. Murray. I was living with my mother. We were in pretty comfortable circumstances. We were living by ourselves. I had never seen anyone with leprosy to my knowledge. I first noticed my nose begin to swell, and afterwards the rest of my face. I was taken to New Somerset Hospital in July, 1889, for what, Dr. Eaton said, was leprosy in my legs. Afterwards to the Old Somerset Hospital; then brought here."

Present Condition, 9th May, 1890.

Symmetrical enlargement of face; right ear more than left; left cheek, large scar of ulceration, which broke out soon after admission, but soon healed; some infiltration and thickening of tongue; fingers of right hand numb, another contracted; left hand comparatively free; some remains of infiltration in forearms, but this has much diminished; some infiltration about the ankles, which are at present suppurating. Infiltration extends nearly as far as knees. Body free; no special indication at seat of vaccination. Has been under treatment since admission. First, iodide and oil gurjun from date of admission, then gurjun alone. General health much improved. Condition of arms and legs much improved.

As copied from case-book.

P. B. TRAVERS STUBBS.

Dr. S. P. Impey, Medical Superintendent, Leper Settlement, Robben Island, South Africa, says:—" I wish to draw your attention to one very serious matter in respect to the spread of leprosy. It is contagious, and can be communicated from one patient to another by inoculation. In South Africa the reprehensible practice of arm-to-arm vaccination is carried on to an enormous extent. I have always held very strong opinions on this subject, and consider that many loathsome diseases are spread by means of the vaccinator's lancet. No medical man would take lymph from a patient in whom the disease is visible; but in how many of these cases is it not latent? For years I have not vaccinated except with animal lymph, and think that some means should be adopted to stop the dangerous practice of vaccinating with humanised lymph; rather allow the patients to have small-pox, where there is a chance of recovery, than force leprosy upon them. It is a noteworthy fact that, since the introduction of the art of vaccination, leprosy is spreading with rapidity. I am a firm believer in the efficacy of vaccine, but consider the arm-to-arm vaccination is a most dangerous practice and one which has led to untold misery."—*Extract from Special Report on Leprosy, from Robben Island, for 1891, in Reports presented to both Houses of Parliament by command of His Excellency the Governor of the Cape of Good Hope, Capetown.*

Dr. Alexander Abercromby, of Cape Colony, writing from Capetown, April 20th, 1892, says that, if a drop of blood gets mixed with the vaccine lymph in the operation of vaccination, then the disease (leprosy) may be transmitted in this way, but he is of opinion that, without the blood, there is no danger. So far as the transference of syphilis and other deadly diseases is concerned, we know that this can be done with lymph of unimpeachable quality and without admixture of blood. In the January (1890) number of the "Archives of Surgery" Mr. Jonathan Hutchinson records several fatal cases, and another fatal case in July, 1890 (p. 23), all following vaccination. Mr. Hutchinson observes:— "There is not the least reason to suspect any want of care in the vaccination or defect in the lymph." On July 6th, 1881, Dr. Robert Cory, Superintendent of the Calf Lymph Department, London, succeeded experimentally in transferring syphilis to himself with lymph free from admixture of blood.

Rev. Canon Baker, formerly Chaplain at the Leper Colony, Robben Island, writes to me, June 1st, 1892:— "I have not met with any medical man who has been any considerable time in this Colony who affirmed that

DR. P. HELLAT'S OPINION.

vaccination, from arm to arm especially, was not attended with danger in the direction of your inquiry."

Dr. P. Hellat, the leader of the movement for stamping out leprosy in the Baltic Provinces, Russia, writes to me from St. Petersburg, May 1st, 1892, that at the time when he published the result of his investigations into the spread of leprosy, he had omitted to take into account "what might be ascribed to the spread of vaccination." The matter "was of great importance, and one that cannot be considered an open question." Dr. Hellat says that in Russia no one is admitted to school without the marks of vaccination. Re-vaccination is resorted to on the occurrence of serious small-pox epidemics.

CHAPTER V.

LEPROSY AND RE-VACCINATION.

THOSE who have studied the literature relating to the remarkable recrudescence of leprosy in tropical countries, or made personal inquiries where the disease is prevalent, must have been struck with the number of cases of persons whose family history has been shown to have been entirely free from all taint of this disease. In some instances the afflicted persons state they have never seen or come in contact with a leper. It is usual amongst the well-to-do classes to keep the malady a profound secret as long as possible; but a time comes when the disease discloses itself to the family physician or casual observer, and the friends are then led to inquire as to how the disease originated. Close inquiries reveal the fact that with them it is not hereditary, and the afflicted member is free from sore or abrasion of the skin. How, then, could the dreadful leprous poison have been contracted? The doctor then suggests the vague and much misunderstood word "contagion," or malaria, or diet, all of which theories are now rejected by leading lepra authorities. The unfortunate patient remembers that during a small-pox scare by the advice of his doctor he was re-vaccinated. Inquiries are made as to the vaccinifer, and it is not seldom discovered that the vaccine virus was taken from a subject in which the leprous taint—then not discernible—has since fully disclosed the ulcerations of unmistakable tubercular leprosy. It will be said that this is a hypothetical case. From much conversation with intelligent observers in many countries where this disease prevails, including both English and native medical practitioners, public vaccinators, and army surgeons, I am convinced that this is by no means an uncommon experience. So strong is the belief in the existence of this danger, that soldiers in our tropical colonies, when subjected to vaccination, often resort to the practice of squeezing out the virus, if they can do so without observation, and use carbonate of soda, borax, and other disinfectants in order to neutralise the effects of the poison. This dread of vaccination is due to the knowledge amongst our troops, often born of bitter experience, that vaccination is not seldom the factor of disgusting and intractable diseases, the fatal cases of which are invariably registered under secondary causes, without disclosing the originating source of the malady. The leprous taint is more common in the army than is generally supposed. In my various inquiries I have heard of several soldiers being so affected. In his work on "Leprosy" (p. 71), Dr. Munro refers to the case reported by Dr. Liveing of a private soldier, a leper, in British India, who had several brothers and sisters all older than himself and all healthy. The same authority cites "the case of an officer mentioned in the Report of the College of Physicians, p. 241. Landré speaks of Dutch private soldiers being affected, while ladies never are, they never being exposed to contagion" (p. 45). If Dr. Munro had explained that ladies are not exposed to compulsory re-vaccination, he would have touched the solution of one source of infection. Dr. Sutherland, of Patra, says:—"When serving with the native army, I found repeatedly that men who had in early life the character which I regard as a proof of the existence of a leprous taint, which I have already described, frequently had to be invalided in after years for leprosy, and subsequent observation and inquiry have led me to the conclusion that the opinion I have formed regarding what I have named a leprous taint was correct, and that this condition precedes the appearance of the disease in its aggravated form; and I think I am warranted in concluding, from the data given above, that this leprous taint exists in one out of every ten of the adult rural population of this district. In stating this, I am aware that my views will probably astonish persons who have not given the subject the attention I have."—*Royal College of Physicians Report, p. 188.*

In a note, entitled, "Leprosy in Livonia," the *British Medical Journal* of August 20, 1887, p. 423, says:—"The disease (leprosy) was introduced by a discharged soldier from Southern Russia." The same journal, in its issue of October 22, 1887, in a note on "The Spread of Leprosy," quotes M. Besnier, a member of the French Academy of Medicine, to the effect that, since the extension of the French Colonial possessions, soldiers, sailors, and missionaries, have fallen victims to leprosy in large numbers. The *British Medical Journal*, January 12, 1889, p. 93, in an article entitled "Transmission of Leprosy," mentions several cases of leprosy, including two young soldiers who became lepers. And the number for March 23, 1889, pp. 668-9, publishes the details of a case of mixed leprosy—tubercular and macular—reported by Dr. O. Caroll to the Royal Academy of Medicine in Ireland. The patient was an army pensioner, who had served in India and the Cape, "but he had not been, so far as he is aware, in contact with any leprous person." Again, on April 13, 1889, the same journal publishes a letter on "Leprosy in the United Kingdom," in which the writer says:—"T. H., after a service of twenty-two years in India under the East India Company, returned home, and soon became a victim to tubercular leprosy." Another case of a soldier, formerly in India, who is now a leper in the wards of a large general hospital, is mentioned by Mr. C. S. Loch in the *British Medical Journal*, July 13, 1889. And in its issue of December 21, 1889, the same journal gives the fatal experience of this re-vaccinated part of the population, observing, on the authority of Dr. Olavide, of Madrid, that in the (leprosy) infected provinces of Spain—Jaén, Cordova, and Guadaljura—"most of the sufferers are missionaries or soldiers."

Madhub Chunder Ghose, Medical Officer in charge of the Leper Asylum, Calcutta, in a communication to the Hon. H. Beverley, dated Calcutta, 27th August, 1889, says :—" It is well known that a distinguished officer of the Indian Medical Service became affected by leprosy during his residence in India."

These cases are too numerous to be explained by the theory of coincidence, and similar instances can be heard of by inquiry in every country where leprosy is endemic. In none of the foregoing instances does it appear that any inquiries were made as to re-vaccination being a possible source of the infection. No doubt the medical practitioners had their private opinions as to the causation of the cases examined, which, however, they have carefully withheld. In one of the cases a press reporter remarked on the reticence of the doctor, who, he said, "absolutely refused to give me any details beyond those which may be said to be available to any member of the public." He admitted, however, that the case was undoubtedly one of leprosy.

Soldiers, it should be observed, are picked men, living amidst healthy surroundings under superior hygienic conditions. They do not associate with lepers, but, on the contrary, carefully shun them. Nor can the disease in the instances I have cited be due to heredity. Re-vaccinated soldiers appear to be more liable to the disease than the European residents in countries where the disease prevails. However much opinions may differ as to other sources of causation, all authorities admit that leprosy is an inoculable disease and communicable by vaccination. Is it not probable, therefore, that the cases I have quoted are due to contamination at the point of the lancet in re-vaccination? Since the terrible disaster at Algiers in December, 1880, when 58 soldiers of the 4th Regiment of Zouaves were syphilised by re-vaccination from a Spanish child "of remarkably healthy appearance," I have made it a practice in all tropical countries which I have visited to interview soldiers and question them on the results of vaccination in the army. I could fill a long chapter of this book with details of disastrous consequences where soldiers have been invalided at hospital with tumours, abscesses, and sometimes intractable maladies directly due to vaccination. Some have had their arms amputated and been discharged from the service, and others have sustained lifelong injuries. It must be noted that soldiers, when quartered in tropical countries where leprosy and syphilis prevail, are often obliged to submit to several re-vaccinations. In Capetown, a soldier belonging to the North Stafford Regiment, on duty near Government House, informed me (February 9th, 1892) that since he joined the army he had been vaccinated thirteen times. He was badly pitted with small-pox.

In an article on "Leprosy in the Baltic Provinces," by Dr. P. Hellat, dated St. Petersburg, in the *Journal of the Leprosy Investigation Committee*, December, 1891, the writer cites the following case which he considers speaks "very strongly for infection":—"A young man coming from a leprous-free district is called for military service, through which he comes to the South of the Empire to a leprous village, gets ill after two years with *lepra tuberosa*, returns home, lives with his mother, who after two years is attacked with *lepra anæsthetica*." Dr. Hellat omits here to note that the first danger to which the recruit is subjected on joining the army is that of being inoculated with contaminated virus in the process of vaccination. All other risks, unless he has sores or wounds on his body (very improbable in a recruit who has recently undergone medical examination), are comparatively infinitesimal. For the serious risks incurred by re-vaccination our troops receive no compensating advantage in the way of immunity from small-pox, either of a sporadic or epidemic nature. According to the army medical reports, we find that among troops in Egypt in 1889 there were 42 cases and 6 deaths from small-pox. In the Bengal army in 1889 there were 71 cases and 8 deaths. In the Bombay contingent there were 49 cases and 3 deaths; and among the troops in Madras 32 cases and 6 deaths. In all, there were, in 1889, among this picked body of healthy re-vaccinated men, 202 cases and 23 deaths. How many cases of erysipelas, eczema, syphilis, tuberculosis, and leprosy were due to re-vaccination is not stated, but we know that after every general vaccination order, a number of soldiers are disabled for a time, some more or less seriously. I produced evidence before the Royal Commission on Vaccination (Vol. III., p. 116) that in 1882 the Federal Government of Switzerland, owing to re-vaccination disasters, rescinded the regulation in the army, and in 1883 M. Weitzel, the Minister of War for Holland, for similar reasons, issued an order to the effect that re-vaccination was no longer obligatory in the army of that country.

CHAPTER VI.

OTHER ALLEGED CAUSES OF LEPROSY.

I CAN only refer very briefly to causes of leprosy other than those detailed in previous chapters. Next in importance to that of contagion is the question of heredity. The theory of heredity, as a factor in the propagation of leprosy, is rejected by leading authorities such as Hansen, Hjort, and Arning, and by many writers it is not even mentioned. Dr. Arning, while failing to discover evidence that the disease is congenital, is of opinion that a certain weakness to resist its attacks may be transmitted.

Dr. G. Armour Hansen, who discovered the *bacillus lepræ* in 1873, has devoted much attention and research to the question of heredity, and failed to find a decided and indisputable proof of its influence in the United States, though hundreds of thousands of immigrants had leprous parents in Norway. — *Minnesota State Board of Health Report, 1889-90, p. 43.*

Dr. W. Munro says :—" I do not deny that leprosy may be occasionally hereditary, but only say that it has never been proved to be so."

Dealing with the subject of the alarming increase of leprosy in Hawaii, the victims amongst a clean nation multiplying faster as the years roll on, until it has invaded nearly every district in the Archipelago, Dr. W. Hillebrand observes :—" And, mark well, in all this, hereditary taint, from the nature of the case, has no share, or, if any, only a most subordinate one."—*Leprosy a Communicable Disease, by Dr. Macnamara, p. 59.*

The *Lancet*, March 31st, 1883, p. 555, observes:— "Morehead, as well as Louis and Cunningham at Almora, came to the conclusion that heredity could not be a great factor in the increase of leprosy in a district, inasmuch as lepers have comparatively small families, who suffer a high rate of mortality, and therefore the survivors are only just numerous enough to replace their defunct progenitors."

Dr. George L. Fitch, in his report addressed to the president and members of the Board of Health, Honolulu, in 1884, says:—" Heredity plays but little figure in the spread of the disease, because we find that, after sending more than 2800 lepers, during a period of eighteen years, to Kalawao Leper Settlement, there are only twenty-six children alive, and only two of these children are lepers."—*Appendix to Report on Leprosy in Hawaii, 1886, p. 31.*

In a note to the Secretary of the State Board of Health, Minnesota, U.S., December, 1888, Dr. Christian Gronvold observes:—" Our experience in the North-West has made it probable that the disease is not hereditary. Not a single case has been discovered, after forty years of immigration, where a child, born in these States of leprous parents, has inherited the disease."

The same authority concludes an article on " Leprosy in Minnesota" as follows:—" I cannot here relate all my observations in detail. I will only tell what I have found in regard to the occurrence, or rather the disappearance, of *lepra* in America (N.W. States). Of about 160 lepers who have immigrated into the three States named (Wisconsin, Iowa, Minnesota), thirteen are alive, whom I have seen myself, and perhaps three or four more. All the others are dead. Of all the descendants of lepers (and that includes the great-grandchildren of some of them), not a single one has become leprous. This is, in short, the result of my investigations."—*Lancet, March 26, 1892.*

Report of the Select Committee on the Spread of Leprosy, Cape of Good Hope, July 15, 1889. Minutes of Evidence :—

The Hon. Dr. Atherstone, M.L.C., who has practised in the Colony fifty years, chiefly in Graham's Town, where he was District Surgeon for twenty-six years, and who has always taken a great interest in the subject, said, in answer to Q. 352 concerning heredity, that "it is the constitutional diathesis, or cachexia, which, I consider, is inherited; not the dormant germs rendering the individual less able to resist its attacks and subsequent reproduction."

In an article on the "Cause of Leprosy," by Sir William Moore, K.C.I.E., Hon. Physician to the Queen, late Surgeon-General with the Government of Bombay, the author says :—" We are told that leprosy is caused by eating fish, and, therefore, cannot possibly be syphilis. Now, I lived and worked many years among the inhabitants of the semi-desert districts of Western India, who never see fish. The sea is hundreds of miles away, and there are no lakes or rivers. Even dried fish did not penetrate into those remote districts. Fish was practically unknown as an article of diet. Yet there is a considerable amount of leprosy in those countries. At recent meetings of the Medical and Physical Society of Bombay, the subject of leprosy was exhaustively discussed, and the members were unanimous in discrediting fish as a cause of leprosy. The members, being both native and European practitioners, private and in the public services, who are constantly seeing leprosy, are perhaps better qualified than any body of men to give an opinion on this matter. In the districts of Western India above referred to, salt is made from the earth at most villages. The people have as much salt as they want. Yet leprosy has been attributed to an absence of salt. It has always been ascribed to a vegetable diet, to new rice, and to diseased grain. But the kind of food does not appear to influence the disease further than, like insanitary conditions, insufficient food, and food deficient in required elements, that it induces a state of constitution rendering the subject more liable to almost any malady. Again, it has been advanced that leprosy is not, like syphilis, contagious. But the communication of leprosy has certainly been proved. Direct proof seems to have been afforded by the experiment on the criminal at Honolulu. In the case of Keanu, *bacilli* were found to have multiplied at the seat of inoculation. . . . Leprosy very frequently commences on the extremities. Natives in India do not generally wear shoes and stockings; their feet are thus very liable to become abraded or wounded; a leper's slippers may easily be taken by mistake, and conveyance of discharge is, I believe, thus frequently accomplished."

In an article in the Fourth Annual Report of the State Board of Health for Massachusetts, U.S., by Dr. Samuel W. Abbott, reprinted in the Hawaiian

"Report on Leprosy in Foreign Countries," 1886, is the following, under the title of "Leprosy in its Relation to Public Health":—" The questions which render leprosy a matter of special interest, as affecting public health, are those of etiology, modes of propagation, and the question of contagion.

"The causes of leprosy have been sought for in the peculiarities of climate, soil, diet, and habits of life. As regards climate and soil, the wide geographical distribution of the disease would seem to preclude them as elements or factors of causation. Opinions differ much as to the question of diet. The eating of tainted fish has been strongly urged as a cause. Leprosy is found in a most aggravated form among fish-eating people, as in Norway and Crete, and, on the other hand, it also prevails in inland districts where fish is but little used.

"Doubtless an improper diet and bad hygienic surroundings aggravate the disease.

"All these causes acting together for centuries did not produce the disease in the Hawaiian Islands, nor was it known until some time after the islands were open to foreign trade and commerce with other nations."

Mr. Jonathan Hutchinson says:—" The suspicion that vaccination has been the means of spreading the disease in the case of the Sandwich Islands has been entertained." The spread of leprosy, according to this distinguished surgeon, is due to eating fish, a theory which finds very few supporters amongst those who have studied the disease in leprous countries. The Parsees are great consumers of fish, and leprosy amongst them is of rare occurrence. With regard to the fish theory, which has obtained currency owing to the eminence of its author, I may observe that in my travels in leprous countries I have hardly met with a single advocate of it, and those most practically conversant with the disease at leper institutions consider it both far-fetched and irrational. In an article on "Leprosy in Kashmir," by Dr. A. Mitra, Chief Medical Officer in that country, in the *American Journal of Medical Sciences*, Philadelphia, pp. 22 and 23, the author says:—"As to Hutchinson's fish theory, the Goojurs do not get fish. Since this theory first came to my notice in the pages of the *Lancet* I have always asked lepers if they have been fish-eating, and in the large majority of instances the reply was in the negative. The theory is untenable in India, where we do not find the disease more prevalent among fish-eating people than among abstainers from such food, as *vaisnabs*. The Kashmiris, among whom leprosy is rare, eat fish, fresh, dried, and salted. In India European sportsmen, planters, etc., use largely preserved fish, but there are no facts to show that fish-eating ever produced leprosy among them. High-class Hindu widows are strictly prohibited from taking fish, but I have seen several cases of leprosy among them. But the fact that the leprosy is common among Goojurs completely disproves the fish theory." In "Notes on Leprosy in Norway," Dr. Hercules MacDonnell observes:—"Nowhere did I perceive that any credence was attached to the fish origin theory. Dr. Kaurin's writings on this special subject are widely known abroad."—*Lancet, August 31, 1889.*

Dr. C. N. Macnamara says:—"Among the Norwegians putrid and dried fish are said to give rise to leprosy;

others fancy that rice prepared in a particular way is at the root of the evil. We may be sure that the theory of bad and salted food being a cause of leprosy does not apply to many parts of India, for the natives, as a rule, do not eat salted food, and certainly are not in the habit of consuming putrid fish."—*Leprosy a Communicable Disease, p. 48.*

In a paper read before the New Brunswick Medical Society, July, 1889, by Dr. Murray MacLaren, M.R.C.S., on "Leprosy in New Brunswick," Mr. Jonathan Hutchinson's Fish and Food theory is dealt with. After quoting Mr. Hutchinson, the writer says:—"This view is not at all borne out by what can be observed in our own affected district, which is only 45 miles in length, and of the 82 cases already mentioned, 58 have arisen in the parish of Tracadie alone, which has been the headquarters of the disease, while the remainder come from the other parishes: Niguac, 3; Pokemonde, 9; Shippegan, 6; Caraquette, 6. It does not seem possible that this district, and especially Tracadie, should have food in any way different from a large part of the extensive northern and eastern coast, which is quite similar to the leprous district in soil, climate, food, including fish, and inhabited by a similar race of people with the same manners of life. Besides this, the fact that no case is known to have occurred among the Indians dwelling within the affected area helps to disprove this theory."—*Maritime Medical News, July, 1890, p. 50.*

Dr. Julius Goldschmidt, in a communication on "The Madeira Leprosy," says:—"Those districts where mostly fish is consumed, and sometimes in an unhealthy state (coast villages), are freer from the disease than in other parts of the island."—*Journal of the Leprosy Investigation Committee, No. 4, December, 1891.*

Dr. Max Sandreczki, Director of the Children's Hospital, Jerusalem, says:—"Leprosy in Palestine is developed by insanitary conditions, of which I will enumerate the chief: impure air; the tainted exhalations which prevail in the villages and unhealthy habitations of the fellaheen; the water supply, often stagnant and deteriorated; the oil and fat (used for food), rancid, or salted beyond measure; olives and cheese in a state of decomposition; meat rotten, or coming from animals diseased, or worse still. Add to all this extreme uncleanliness, the utter absence of skin action, and then one may easily explain derangements in the tissues of the skin, in the lymphatic and ganglionic systems, and, in a word, complete disorder of all the nutritive functions."—*The Lancet, August 31st, 1889.*

The Hon. David Dayton, President of the Board of Health, Honolulu, says:—"I am not of the opinion that leprosy is always hereditary, so many cases proving to the contrary. By referring to tables in Mr. C. B. Reynolds' reports, it will be seen that a large proportion of the girls in the Kapeolani Home were children of leprous parents without becoming diseased themselves." —*Report, Board of Health, Honolulu, 1892, p. 42*

CHAPTER VII.

INADEQUACY OF MEDICAL THEORIES OF CAUSATION.

IN conversation with lepra specialists in countries where leprosy is endemic, nothing is more common than the admission that the alleged exciting causes of the disease, such as contagion, heredity, or malaria, are quite inadequate to account for the rapid progress of leprosy over a wide area in recent years. Dr. Alzevedo Lima, chief medical officer of the hospital, Rio de Janeiro, observes: —" Even at the present time, in spite of the progress in sanitation, and the more favourable conditions of life among the Brazilians, there are still centres of contagion, whence we receive patients for the hospital in larger numbers than from other places. Throughout the whole country, in fact, this terrible disease is still rife. This being so, if we consider the differences in climate, the food, the habits of life, the sanitary conditions, etc., of the various regions of this vast country, it is difficult to believe that the etiological factors still given as determining causes of the disease can be in themselves a sufficient explanation. At the most, they can only act as favouring conditions, and explain the greater or less frequency of the malady, its predilection for certain places or for certain classes of the people."—*Journal of the Leprosy Investigation Committee, December, 1891, p. 24.*

Alluding to this subject, Dr. Alzevedo Lima, writing to me from Rio de Janeiro, May 20th, 1892, says:— "None of the ordinary etiological factors explains satisfactorily the spread of leprosy in this country, where one finds focuses in places altogether different in the climate-telluric conditions, and where, besides the sick people who live in bad conditions of feeding, etc., there are many others who live in luxury and belong and dwell amongst the best society. The inheritance also does not account for it in a very large scale. About this last point I always take very great care to inquire of the sick people who are taken to the Hospital dos Lazaros, where we have had 242 patients during these fourteen years past that it has been under my direction. . . ." Referring to the inoculation of leprosy by means of vaccine virus, Dr. Lima says:—"I believe in the possibility of the fact, not seldom, and I suspect this may have been the cause of the spread of the disease amongst our people, specially in the country, where, with the absence of a doctor, the vaccination is done by someone not professional, and therefore incapable of distinguishing whether they are using the vaccine taken from a pure source or not. Moreover, to those who know how long it takes for the incubation of leprosy, and how difficult it is to diagnose it in its initial stages and in several of its forms and varieties, it is easy to know that even the professional man may have given rise to the spread of the disease in that way.

"Amongst 62 persons affected with this disease now in treatment in the Hospital dos Lazaros, in this city, 26 were vaccinated, being 10 in Rio de Janeiro, 8 in the

State of St. Paul, 6 in the State of Minas, and 2 in Portugal; the remaining 36 have never been vaccinated. Abstracting the 10 of Rio de Janeiro, where the vaccination is gone through with every care and the vaccine taken directly from the cow, we have 8 come from St. Paul, either from the country or from small villages where the vaccination is performed, as a rule, by people not competent, and with the lymph taken at random from any person amongst their own people, where leprosy prevails endemically. In any of these States, in the absence of a physician, any clever man undertakes the duty of vaccinating amongst the people, and this may very likely be one way of propagating the leprosy."

Dr. Alexander Abercromby, of Cape Colony, referring to the causation of leprosy, says:—" It is evidently dependent on some vitiated state of the blood, and that acquired in many instances, as has been clearly ascertained, by hereditary predisposition. In many cases, however, it occurs where no such predisposition can be traced, and in persons whose parents were perfectly healthy, and who evinced during their lives no trace of the disease whatever. In these cases we are led, therefore, to seek for other causes to account for it."—*Thesis on Tubercular Leprosy, p. 15.*

In the Report on the Annual Returns of the Civic Hospitals and Dispensaries in Madras for 1888, p. 14, under the head of Vizagapatam, Surgeon-Major Sturmer says that leprosy is on the increase in the district, and observes:—" This year I have seen many fresh cases of leprosy in adults as well as in children, in whom no hereditary taint could be traced. They evidently had contracted the disease from some outside source, for in each case it was ascertained that no other member of the family was affected."

The " Report on Leprosy in New South Wales," May 13th, 1891, forcibly exhibits the unsatisfactory condition of modern inquiry into the etiology of the disease. No mention is made as to the chief factor of causation.

The Secretary of the Board of Health in New South Wales, Mr. Henry Sager, observes :—" The detailed history of the cases given in Appendix C., though of very considerable interest, does not furnish any grounds for definite conclusions as to the causation and spread of the disease. There are no data on which to advance a view of spontaneous, climatic, dietetic, mal-hygienic, or hereditary origin of the malady, and nothing of scientific accuracy to be adduced as to contagion, though the evidence in several cases points more or less strongly in this direction."—*Journal of the Leprosy Investigation Committee, No. 4, December, 1891, p. 50.*

The Report of the Inspector of Asylums for 1890, presented to both Houses of Parliament, Cape of Good Hope, under the head of " Female Leper Wards," p. 32, says :—" The fact which stands out most prominently in making these records is the absence of any history of direct contagion, or even with contact of a known source of the disease in almost all of the cases which have been investigated."

When, about a year later, Dr. S. P. Impey, the present medical superintendent of the asylum, Robben Island, unable to account for the spread of leprosy by popular medical theories, began a more careful and exhaustive investigation than his predecessors had ventured upon, and included vaccination, hitherto ignored, as one of the possible factors, he had no difficulty in tracing a number of cases directly to this source. Dr. Impey felt it his duty to warn the Government of this danger.

In concluding a second paper on "Notes on Acquired Leprosy as observed in England," by Mr. Jonathan Hutchinson, in the *British Medical Journal*, July 6, 1889, this distinguished pathologist observes:—

"The twelve cases which I have quoted do not, I believe, comprise by any means all the examples of leprosy beginning in patients of British birth and descent which have come under my observation. They are all, however, of which I am able at present to find record in my note-books. One and the same criticism may be said to apply to all. They are the examples of the acquisition of a specific disease by healthy persons who had no inherited predisposition. In no single instance had the person so acquiring it been exposed to any degree of hardship, or deviated in any definite manner from the ordinary conditions of a cleanly and well-regulated life. In every case the acquisition of the disease had occurred in some country where it was known to be prevalent, the East and the West Indies being the chief localities. I submit —as, indeed, I have already suggested—that there are only two suppositions open to us by which to explain the *de novo* acquisition of such a disease in cases such as these. The patients must either have received the specific contagion of leprosy on some part of the skin or mucous membrane, or they must have swallowed it in connection with food. Both these suppositions are possible. I may confess, however, that to my own mind one of them seems far more probable than the other. In no single instance had there been any known exposure to contagion. In no case had the patient associated with anyone suffering from leprosy, and in most instances the statement given was that they had but rarely seen lepers, and had certainly never come near them."

Amongst the possible sources of leprous contamination, Mr. Hutchinson, while inclining to what is known as "the fish theory," mentions "the perils of vaccination."

CHAPTER VIII.

LEPROSY AND VACCINATION AT THE INTERNATIONAL HYGIENIC CONGRESS.

IN the month of June, 1890, I presented before the Royal Commission on Vaccination certain facts tending to show the increase of leprosy in various countries which I had visited, and that this increase was largely due to vaccination. In the following autumn I extended my inquiries to other countries, and discovered the same sinister results; and in the following April I published my conclusions in a pamphlet entitled "Leprosy and Vaccination," which has been widely circulated in the countries where the alleged cases of leprous invaccination have occurred. But no attempt was made to reply to the allegations until August, 1891, when Dr. Phineas S. Abraham, the Secretary of the Leprosy Investigation Committee, read a paper on the subject before the International Congress of Hygiene and Demography in London. The following is the report of his address, from the *Lancet* of August 22, 1891 :—

"*On the Alleged Connection of Vaccination with Leprosy.*— Phineas S. Abraham, M.D., read a paper on this subject. Accepting the bacillary theory of leprosy, and believing that instances have occasionally been reasonably demonstrated of its communication from one infected person to another previously healthy, Dr. Abraham has sought for evidence as to every possible means of inoculation, vaccination included. In this paper he gave a short account of the principal inquiries bearing upon the subject, and discussed the facts which have been alleged to connect leprosy with vaccination. The statements, or supposed cases, of the following observers were among others alluded to :—Sir Erasmus Wilson, Sir R. Martin, Dr. Bakewell, Dr. Tilbury Fox, Dr. Castor, Dr. Ebden, Mr. Malcolm Morris, Dr. Bemiss, Dr. Hildebrandt, Dr. Arning, Dr. Mouritz, Dr. Rake, Archdeacon Wright, Professor Gairdner, Dr. Black, Drs. Swift and Montgomery, Surgeon Brunt Dr. Piffard, Mr. Hillis, and Professor Leloir. It was pointed out that the question had been in the minds of inquirers for many years past, and that the supposed instances brought forward were comparatively few. Even the most suspicious cases, such as those adduced by Professor Gairdner, Mr. Hillis, and others, were open to the objection that there was nothing to show that the subjects had never been exposed to any other possible means of inoculation or contagion, had never been in contact with lepers, or had never had to do with food or anything else which might have become contaminated by lepers; in short, we could not be sure that, having been born, or having lived for some time in a leper land, they had not been exposed to other pathogenic conditions of the disease. Dr. Arning's and some of the other observations were quoted in full because they had been much twisted, and false deductions have been drawn from them. The evidence from Scandinavia was significant; vaccination had been compulsory in Norway for many years, and largely practised from arm to arm in the leprous districts, and, as Dr. Hansen stated, no case of transferring leprosy therefrom had been hitherto known. Leprosy, indeed, was there steadily decreasing. In China, according to Dr. Manson, leprosy was common in the district where vaccination had been practised for the last sixty or seventy years; but, on the other hand, it was more common in districts where vaccination had only been recently introduced, and was practised to a very limited extent only. Some fresh facts and definite information on the subject were hoped for from Mr. William Tebb's late evidence before the Vaccination Commission. A great deal was, of course, made of the observations of Professor Gairdner, Dr. Arning, Dr. Castor, Mr. Hillis, and of some of the others alluded to above. One of them held the view that the spread of leprosy might be due to *syphilitic* vaccine lymph, and another had written a pamphlet on the subject which was full of inaccuracies. In point of fact, although *a priori* the possibility of an occasional accidental inoculation of the disease by vaccination might be admitted, up to the present time no absolutely clear and incontrovertible evidence connecting vaccination with leprosy had been forthcoming; and, in Dr. Abraham's opinion, anyone who said that vaccination was to any extent responsible for the spread of leprosy talked arrant nonsense. Nevertheless, from what was known concerning the introduction of bacillary diseases in man and animals, it certainly behoved medical men to be extremely careful in the selection of their lymph for vaccination; and in a country where leprosy was rife it seemed to him that it would be advisable to exercise particular caution, and, if possible, avoid, as was now being done in Hawaii, an indiscriminate arm-to-arm vaccination among the natives. The question of the possibility of transmitting leprosy bacilli by vaccine is receiving attention on the part of the Indian Leprosy Commission, and a paper on the subject by Drs. Bevan Rake and Buckmaster will appear in the next number of the *Leprosy Journal.*"

In answer to this statement I sent the following to the *Lancet:*—

LEPROSY AND VACCINATION.

(To the Editor of the *Lancet.*)

SIR,—Dr. P. Abraham's address before the recent International Congress of Hygiene and Demography on the alleged connection of vaccination with leprosy, reported in the *Lancet* of 22nd August, is hardly calculated to allay public anxiety in our tropical colonies where the disease is endemic. As he has referred to the evidence which I gave on this subject before the Royal Commission, I shall be glad if you will kindly allow me a short space for explanation. Dr. Abraham has furnished a tolerably large list of authorities who have either pointed out this particular danger or supplied particulars of cases where leprosy has in their opinion been invaccinated, but he adds that these cases are comparatively few. Allow me to observe that Dr. Abraham has omitted to mention that, of the distinguished medical witnesses, some have adduced several cases of invaccinated leprosy, and others refer to a "prolific," "serious," "alarming" increase of leprosy due to vaccination. Some of the cases are introduced with reluctance by practitioners, who know the damaging effect of these allegations upon a prescription lauded as "the greatest discovery in the history of medicine." Dr. Abraham quotes Dr. Hansen as stating that in Norway no case had been hitherto known of the communication of leprosy by vaccination. Permit me to say that this well-known pathologist has given his emphatic opinion, which I quoted before the Royal Vaccination Commission, that the chief means of disseminating leprosy is by inoculation, and that in Norway the greatest possible care is observed to prevent lymph being taken from leprous subjects. Dr. Abraham advises medical men to be "extremely careful" in the selection of their lymph, especially in countries where leprosy is rife, and to avoid arm-to-arm vaccination, "as is now being done in Hawaii." I beg to state that less than a year ago, as I know from personal inquiry, the bulk of the vaccinations in Hawaii were performed with humanised virus, it having been found that animal lymph produced excessive inflammation and many terrible cases of ulceration. In Ceylon, where leprosy is endemic, and, according to Dr. Kynsey, the Surgeon-General, increasing from some occult cause, arm-to-arm vaccination is principally in vogue, as it is also in the West Indies. If leprosy is spread chiefly by inoculation (and this source of infection is more generally accepted amongst dermatologists than any other), there is no mode of inoculation so widely prevalent as vaccination ; and, having investigated the subject in all quarters of the globe, I attribute largely to this cause the alarming recrudescence of the disease which has taken place during the last thirty years in our colonies and most other leprous countries.

WILLIAM TEBB.

Devonshire Club, St. James's,
London, 5th September, 1891.

This letter was refused insertion.

It is clear by Dr. Abraham's mode of argument that leprous vaccination, as an important factor in the increase of the disease, is the last thing he will admit. All other possible sources of dissemination must be excluded before a theory so fatal to medical prestige can be tolerated. Any other of the numerous theories promulgated to account for new centres of leprous contamination the doctor is ready to consider, but vaccination (to use a classic phrase) "must be preserved from reproach," and the reputations of its distinguished advocates maintained. His opinions, however, do not seem to meet universal approbation, even amongst the medical profession. Commenting upon Dr. Abraham's address, the *Hospital Gazette*, London, of August 22nd, 1891, observes :—" Dr. Abraham has gone to a great deal of trouble to prove, or attempt to prove, that, though leprosy is probably as susceptible of being conveyed by vaccination as is syphilis, there is no well-authenticated case of the kind on record. It is admitted that quite a number of suspicious instances have been reported by competent observers, but Dr. Abraham rules them all out of court on the ground that other possible sources of infection have not been eliminated. This seems to be asking too much. Leprous vaccine is obviously only obtainable in countries where lepers are common, so that, theoretically, the victims must necessarily have been exposed to the special pathogenic influences. It has, we believe, been scientifically demonstrated that the disease can be conveyed by inoculation, and we shall require something more than ths specious special pleading of Dr. Abraham before acquiescing in his conclusions."

CHAPTER IX.

VACCINATION IGNORED IN OFFICIAL LEPROSY REPORTS.

Now that evidence is accumulating in all directions regarding vaccine virus as a propagator of leprosy, attempts are made to minimise the effects of this evidence on public opinion, by alleging that the instances of such infection are few in number, and of no account when put in the scales against the enormous benefits arising out of the application of Jenner's great discovery. This is not the place to enter into the *pros* and *cons* of this much-vexed question, but those who wish to study the facts may do so advantageously in the reports of the Royal Commission on vaccination; in the article on vaccination by Dr. Charles Creighton in the ninth edition of the "Encyclopædia Britannica"; in the able monograph, "Jenner and Vaccination," by the same author, and in Professor Crookshank's instructive treatise, "The History and Pathology of Vaccination." My own views on this subject—the results of a lengthened experience—may be found in the *Westminster Review* for December, 1888, and January, 1889. While I do not admit with Dr. P. Abraham, Dr. George Thin, Dr. Bevan Rake, and others, that the cases of leprosy due to vaccination are few in number, it should be borne in mind that the subject has never been submitted to searching and impartial investigation. In 1862, by request of the Government, a Committee of the College of Physicians prepared a series of seventeen interrogatories which were sent to lepra specialists in all parts of the world, but all reference to vaccination as a possible or probable factor was strictly excluded.

In consequence of the rumours of the spread of leprosy, by means of vaccination, in the Island of Trinidad, Governor J. R. Longden felt it his duty to call the attention of the English Government to the subject, and on the 4th March, 1871, he addressed a dispatch to the Right Hon. the Earl of Kimberley, then Secretary of State to the Colonies. In confirmation of this serious charge against vaccination Governor Longden referred to the report of Dr. Bakewell, the Vaccinator-General of Trinidad, and to certain cases of invaccinated leprosy, and to Sir Ranald Martin and other eminent physicians as authorities for his statements, and added, "This part of Dr. Bakewell's report appears to me to be deserving of your Lordship's attention in connection with the increase of leprosy, which I fear must be admitted to have taken place in the last few years." In paragraph 13 of the dispatch Governor Longden says: "The danger of introducing disease into the system of a previously healthy child by vaccination is possibly a real one, and it is very important, as regards tropical colonies at least, that it should receive the attention of the medical profession." In paragraph 14, Governor Longden pertinently observes that the special danger of spreading broadcast the seeds of leprosy would be worse than the perpetuation of small-pox. Governor Longden's dispatch was referred by the Colonial Department to the Royal College of

Physicians, who were anxious to get rid of the most damaging indictment yet preferred against vaccination. This they sought to accomplish, as far as possible, by ignoring it altogether.

That learned body nominated Dr. Gavin Milroy, F.R.C.S., who arrived at Georgetown, British Guiana, on the 22nd July, 1871, and prosecuted inquiries mainly as to the contagiousness of leprosy in that Colony, and on the 17th October, 1871, he submitted to the Earl of Kimberley a report confirming the conclusion of the College of Physicians, as given in p. xix. of their report as follows :—" The all but unanimous conclusion of the most experienced observers in different parts of the world is quite opposed to the belief that leprosy is contagious or communicable by proximity or contact with the diseased." Dr. Milroy adds:—" My personal observations and inquiries in the Colony all tend in the strongest manner to the same result." No question as to vaccination was submitted for consideration. On the 25th October, 1871, Dr. Milroy reached Barbados, and from thence proceeded to Antigua, and later on visited Trinidad, Dominica, and Jamaica. On reaching Trinidad in November, 1871, he discovered that Dr. R. H. Bakewell in his report had given " countenance to the popular belief as to the transmissibility of leprosy by vaccination," and, with the consent of Governor Longden, inserted an additional question to the Interrogatories for circulation in that Colony. This question, then submitted for the first time, elicited answers abundantly confirming Dr. Bakewell's contention. These answers will be found cited in the communication from Mr. Alexander Henry, and elsewhere, in this volume.

Dr. Milroy's report on this part of the subject is mainly directed to answer and, if possible, disprove Dr. Bakewell's allegation, which, if unanswered, might prejudice the continuance of vaccination. Dr. Milroy says, "What is contended for is, that pure, genuine, vaccine virus, unmixed with blood, cannot be the medium of any contagion but cow-pox:" conditions, as every public vaccinator knows, impossible of fulfilment. At this period the transference of syphilis by means of vaccination was publicly acknowledged, but Dr. de Verteuil of Trinidad, in his reply to Dr. Bakewell, observes: "It is an illogical deduction that, because syphilis is inoculable, leprosy is, or might be, inoculable, the diseases being essentially different." Dr. Browne of Barbados, one of the witnesses, has misgivings on the point, and writes to Governor Rawson, November 8th, 1871, as follows: "It has been a general rule not to vaccinate from the apparently unhealthy, or those of leprous taint, not so much from any opinion founded on fact of the possibility of conveying disease, as from a respect for the general prejudices prevailing." Public opinion, as shown by Dr. Bakewell's evidence, was even at that time in advance of medical opinion regarding the danger attending vaccination. Dr. Reade, the Colonial Surgeon of Singapore, cautiously observes (Dr. Gavin Milroy's report, p. 36): "There is a possibility that the disease (leprosy) may be transmitted from children hereditarily tainted with leprosy, and I strongly advise the continuance of importing lymph from England by every mail, and carefully selecting only healthy children as vaccinifers."

The testimonies elicited by this inquiry as to the communicability of leprosy and syphilis seem to have had no practical effect on the College of Physicians. In a letter from that body on "Vaccination and Leprosy," dated London, August 17, 1871, and addressed to the Earl of Kimberley, it is stated that, while it is admitted that in a few instances syphilis has been transmitted by vaccination, "yet with reference to leprosy it must be observed that there is no evidence adduced beyond the merest presumption that this disease has ever been transmitted by vaccination." And so far from cautioning the public against this fearful danger, and petitioning Parliament to repeal a law which had been productive of so much mischief, the College of Physicians reaffirmed their belief in the benign character of vaccination, and declared "that they cannot press too strongly on your lordship the importance of enforcing the practice of vaccination for the protection of those who are too ignorant to protect themselves, and it would be a grievous wrong to forego so great a public benefit on the mere speculative grounds advanced by Dr. Bakewell."—*Report on Leprosy and Yaws, p. 86.*

The next inquiry emanated from the India office, and is entitled "Scheme for obtaining a better knowledge of the endemic skin diseases of India," prepared by Tilbury Fox, M.D., F.R.C.P., and T. Farquhar, M.D., 1872. Two chapters in the report of this inquiry are devoted to leprosy. The authors here furnish a list of twelve questions for elucidation in regard to the presence and cause of leprosy in different districts, and three questions for leprosy in individuals. In none of these is either inoculation or vaccination specified.

The alarming increase of leprosy in Hawaii, which took place after the introduction of vaccination by the
17

missionaries, once more called public attention to the subject, not only in that group of islands, but throughout the civilised world. In 1885 the Department of Foreign Affairs in Honolulu instituted the most extensive inquiry made up to that time into the causation of leprosy, and the means of its treatment and prevention. The deplorable position of affairs is briefly stated in the introduction to the official report, entitled, " Leprosy in Foreign Countries : Summary of Reports furnished by Foreign Governments to His Hawaiian Majesty's Authorities, as to the prevalence of Leprosy in India and other Countries, and the measures adopted for the social and medical treatment of persons afflicted with the disease." . . . " It is about thirty years since leprosy first attracted any serious attention in the Hawaiian Islands. In the year 1866 the dread disease had gained such a deadly hold upon the native race, that the Hawaiian Government began to attempt to stamp out the scourge by segregation, for it had become a contest for the preservation or destruction of the aboriginal race. To judge by the number of cases in proportion to the population, the disease appears to be more virulent and malignant in the Hawaiian Archipelago than elsewhere on the face of the globe. What has been attempted and accomplished in this twenty years' struggle with a great national calamity appears elsewhere."

" His Hawaiian Majesty's Government being anxious to provide every possible means for the treatment and understanding of the fearful malady, His Excellency Walter M. Gibson, His Majesty's Minister of Foreign Affairs and President of the Board of Health, addressed letters of inquiry to the Secretary of Legation at Ceylon and to the diplomatic and consular representatives of the Hawaiian kingdom in various parts of the world where leprosy was known to exist, making inquiry in respect to the character and treatment of the disease." It is stated that the response to these inquiries has been most generous, more especially from governments of dependencies of Her Majesty Queen Victoria. This interesting document gives reports from every section of the vast Empire of India and its dependencies, from Ceylon, Hong Kong, Siam, the Netherlands and their colonies, the Canary Islands, Norway, Spain, Mexico, Chili, and Guatemala, and an extremely interesting and valuable report from the famous Leper Institution of Tracadie, New Brunswick, Canada. The report from the Secretary to the Government of India being so comprehensive and voluminous, it has been considered expedient to separate it from the other reports.

"In grateful recognition of the sympathy of other afflicted nations, this collection of reports, together with the sad history of its own affliction, is presented to the world by the Hawaiian Government in the devout hope that the Almighty, in his great mercy, may ere long permit suffering humanity to find the means of mitigating the terrible scourge."

From this little known compilation I have made extracts on various matters dealt with in this volume.

In this important inquiry, although at that time the facts had become known throughout the Hawaiian Archipelago as to the spread of leprosy by vaccination, yet such was the reluctance to bring so grave a charge against a practice proclaimed far and wide as "the greatest discovery in the history of medicine," that it was thought expedient to make no mention of vaccination in the interrogatories, and to specify only those points drawn up by the Committee of the College of Physicians in 1862.

Owing to the increase of leprosy in South Africa, inquiries by Select Committees of the Legislative Assembly were instituted in 1883 and 1889, but the interrogatories relate chiefly to the spread of the disease, and to its contagious or non-contagious character. The questions were submitted *vivâ voce*, and vaccination as a possible factor in the dissemination of leprosy is carefully ignored.

An important report by Dr. S. P. Impey, Medical Superintendent of the Leper Settlement, Robben Island, on the dangers of spreading leprosy by means of vaccination, was in June, 1891, sent to the Colonial Office, Capetown. Of this report I have attempted, by repeated personal applications at the Colonial Office, Capetown, and at the Stationery Department of the colony, to obtain a copy, but without success.[1]

In the "Report of the Leprosy Inquiry Commission" for the Colony of Mauritius, published 26th October, 1888, I find the following request in the circular of instructions sent to medical practitioners :—"We would be glad to obtain any facts bearing upon the question of its heredity or contagiousness, upon the conditions favouring its diffusion, upon its treatment, and, finally, upon the best means of preventing its spread in the community.—T. LOVELL, Chief Medical Officer," etc.

---

[1] An extract from this report, through the intervention of a correspondent in South Africa, has since been obtained.

Nothing is said as to the inoculability of the disease, nor is there any inquiry suggested as to vaccination, which has been the means of spreading leprosy in this colony.

The Royal College of Physicians published another report on leprosy in 1889, concerning which the *Lancet*, April 20th, 1889, says:—" A report from the Leprosy Committee was read. It stated that the documents forwarded by the Government on the subject of leprosy since 1887 did not contain much, if any, new information. In view of the fact that there is increasing evidence respecting the communicability of leprosy, the committee repeated 'with greater urgency' the recommendation made in 1887, that the Government should institute a full and careful scientific investigation, which would entail expense and require considerable time. The adoption of the report was moved by Dr. Symes Thompson, who said that the disease was spreading very much among communities in South Africa, and was seconded by Dr. Handfield Jones, who thought that the College should express more definitely its opinion of the contagiousness of the disease and the need for compulsory segregation. This view was not accepted by other speakers, but all concurred in the urgency for a thorough scientific investigation, and it was referred to the committee to draw up a statement respecting the scope of such inquiry, for submission to the Government."

This report was founded upon official and other documents collected up to that date and forwarded by the Government to the College of Physicians. Of these documents I have been unable to obtain copies or even permission to inspect them. The report is signed, "James Risdon Bennett, Chairman," and is dated April 5th, 1889. It was then four years since the disastrous effects of vaccination in Hawaii had been published by Dr. Edward Arning, and two years since Professor Gairdner's remarkable cases of invaccinated leprosy had been made known in the *British Medical Journal.* These and other facts showing the danger of invaccinating leprosy had been laid before the Dermatological Congress of Europe by eminent specialists. And soon after the publication of Dr. Gairdner's cases, Sir William Robinson, Governor of Trinidad, issued a confidential circular to about 30 medical practitioners of that island, containing the question as to whether leprosy is communicable by vaccination, "lymph from healthy vesicles alone being used." Dr. A. S. Black, a well known practitioner, gave particulars of several cases in his own experience, and stated that leprosy was increasing in the island. In his report to the Surgeon General for 1890, p. 34, Dr. Bevan Rake says:—"Some thirty or more Trinidad doctors to whom the same circular was addressed returned negative replies." Dr. Rake omits to state that those who doubt or deny the risk qualify their answers by remarks such as that there is no danger "if pure lymph only is used, and precautions taken in the selection of the vaccinifer and the examination of the pustule," "if there be no admixture of blood" and "healthy lymph is used," "provided the lymph is clear" and "the vaccinifer is free from hereditary taint," "if bovine virus is selected," "with perfect cleanliness of the lancet," etc. Dr. Woodlock mentions that he takes the precautions of constantly importing fresh certified lymph from England. Dr. D. de Montbruñ "dreads vaccination on the ground that syphilis and other cutaneous diseases have been transmitted by it." He also states that nearly all the families in Trinidad strongly object to vaccination with lymph taken from the children of the island, from fear that leprosy may be thereby communicated. Dr. Chitterton says:—"Vaccination is performed in Trinidad in a very unsatisfactory way." It is obvious that the value of the answers is seriously vitiated by the form in which the question is worded. The use of clean lancets and healthy vaccinifers without hereditary taint, however much insisted upon, cannot be made compulsory; and the people are obliged under severe penalties to submit to whatever vaccination is offered, which is chiefly of the leprous and syphilitic variety, collected from miscellaneous native vaccinifers by perfunctory public vaccinators. Indeed, as I have found by personal inquiries in the West Indies, South Africa, and Hawaii, all the precautions admitted to be indispensable for the safe performance of the official rite are habitually disregarded.

The latest inquiry is due to the extraordinary amount of public interest awakened by the published reports of the labours, devotion, and death of the late Father Damien in the Sandwich Islands, and to the accumulation of evidence from many English and French Colonies showing conclusively the increase of this frightful malady. The first meeting was held at Marlborough House, June 17th, 1889, under the presidency of H.R.H. The Prince of Wales; and a dinner in aid of the National Leprosy Fund was held at the Hotel Metropole, London, on the 13th January, 1890, at which also the Prince of Wales presided. A highly influential Committee was subsequently appointed, with Dr. Phineas S. Abraham as secretary. The following letter, with a view of eliciting suggestions, was published in the *Lancet* of the 31st May, 1890, also in the *British Medical Journal*, and in the first number of the *Journal of the Leprosy Investigation Committee*:—

SIR,—With the object of eliciting by correspondence as much information as possible on the subject of leprosy, it is proposed as a preliminary investigation to address a series of questions to the officers of the various leper asylums and to others who may be able to throw some light upon the matter. I am requested to ask you to allow me to invite the co-operation in this inquiry of those of your readers who, from their knowledge of the disease, may be in a position to offer suggestions as to matters of inquiry and as to points of elucidation. Any observations with which the Committee may be favoured will be gladly received and incorporated in the "Journal of the Leprosy Investigation Committee," of which the first number will be shortly published.—I am, sir, yours faithfully,

P. S. ABRAHAM, Med. Sec.

May 26th, 1890.

Dr. Abraham further explains that "although all information bearing on leprosy will be deemed of interest, it will be desirable for observers to direct particular attention to questions relating to the cause or causes and propagation of the disease, as well as to those referring to remedial measures." "It is to be noted that this 'Journal' is published for the purpose of obtaining reliable scientific information on the subject of leprosy, and that it will not be carried on in the interest of any one particular theory. Views from all sides will be admitted; and that the truth may be arrived at, full and free discussion is invited."

Nothing could be fairer, more explicit, or more promising for establishing public confidence and support than these announcements. Realising the importance of the subject and the limitations laid down by Dr. Abraham, I addressed to him, as Secretary of this Inquiry, the following brief communication on the principal point referred to—the causation of leprosy :—

> Rede Hall, Burstow, near Horley, Surrey,
> June 10th, 1890.
>
> SIR,—Observing your note in the *Lancet* of the 31st ult., requesting suggestions as to methods of inquiry, and as to points of elucidation, with regard to the remarkable spread of leprosy, I beg to point out that amongst the questions which it is proposed by your committee to issue to the superintendents of Leper Hospitals, dermatologists, and others, that of the connection of the disease with vaccination should be included. That there is a connection is now admitted by some of the most eminent authorities of the day, including Professor W. T. Gairdner, Dr. Liveing, Sir Morell Mackenzie, Dr. John D. Hillis, Dr. Edward Arning, Dr. Armaur Hansen, and others. Some of these writers admit that not only is leprosy communicable with the vaccine virus, but that new centres of contagion of this hideous disease have been created by vaccination, with most disastrous and far-reaching consequences. Trusting that this important feature of the question will not be overlooked by your committee, and awaiting the favour of a reply, I am, sir, yours faithfully,
>
> WILLIAM TEBB.
>
> Dr. Phineas Abraham,
> National Leprosy Fund, Adam Street, W.C.

To this letter the following reply was received :—

> 2 Henrietta Street, Cavendish Square, W.
> July 2nd, 1890.
>
> DEAR SIR,—I must apologise for not answering your letter before this. With regard to the alleged connection of vaccination

with leprosy, this question will certainly be one of the points to which special attention will be directed on the part of the committee, and an attempt will be made to sift the evidence in an mpartial manner.—I am, dear sir, yours faithfully,

PHIN. S. ABRAHAM, M.D.
W. Tebb, Esq.

This correspondence, implicating vaccination, and pointing out a much-neglected source of danger, is absolutely ignored in the *Journal of the Leprosy Investigation Committee;* while communications from all parts of the world, in which the most diverse and conflicting theories are advanced by persons whose opportunities for observation and inquiry have necessarily been of a very meagre description, have found insertion in its pages.

In a notice of a recent able work, "Leprosy," by G. Thin, M.D. (London : Percival & Co.), in No. 4 of the *Journal of the Leprosy Investigation Committee*, p. 71, no mention is made of the cases of invaccinated leprosy introduced by the author, which occupy several pages of the work. On the other hand, three articles against the theory that leprosy is spread by vaccination are inserted. It is also to be observed that, while numerous communications, pointing out the dangers of vaccination in countries where leprosy is prevalent, have, since the Leprosy Commission was appointed, appeared both in the home and colonial press (some of the writers citing cases of the disease disseminated in this way, and others furnishing the results of their painstaking investigations), no notice of such communications has been taken in the *Journal of the Leprosy Investigation Committee.* Three pamphlets dealing with the subject

—two of them by medical writers—have been treated in a similar fashion. What amount of confidence the public, who have subscribed largely to the National Leprosy Fund, will place on an inquiry so manifestly one-sided, remains to be seen. It is certain that those who have looked for the impartial treatment of this serious phase of the question at the hands of the Leprosy Investigation Committee will be grievously disappointed. Counsel holding a brief for the perpetuation of the Jennerian *cultus* could hardly have exhibited a less judicial attitude than is disclosed by the official documents relating to this latest leprosy inquiry.

CHAPTER X.

OFFICIAL STATISTICS.

THE LEPER CENSUS IN INDIA, 1881, 1891.

MR. J. A. BAINES, Census Commissioner, India Office, has been kind enough to furnish me with the following Table showing the results of the Census in relation to Leprosy in the decades 1881 and 1891.

PROVINCE.	1881.		1891.	
	Males	Females.	Males.	Females.
1. Ajmír,	23	6	20	7
2. Assam,	2,409	906	5,128	1,599
3. Bengal,	40,484	13,490	32,957	11,029
4. Berar,	2,971	777	2,886	1,624
5. Bombay,	7,259	2,559	7,558	2,419
6. Sindh,	166	111	125	84
7. Aden,	0	0	1	0
8. Upper Burmah,	0	0	2,262	1,242
9. Lower Burmah,	2,009	580	2,281	679
10. Central Provinces,	4,430	2,013	3,575	1,780
11. Coorg,	25	18	12	11
12. Madras and Small Feudatories,	10,329	3,846	9,455	3,182
13. N. O. Prov. and Oudh,	14,453	3,369	14,114	2,957
14. Punjaub,	5,333	1,547	3,322	1,029
15. Quettah,	0	0	2	0
16. Andaman,	30	0	1	0
17. Hyderabad,	2,117	872	2,261	716
	92,038	30,094	85,960	28,358

OFFICIAL STATISTICS.

PROVINCE.	1881.		1891.	
	Males.	Females.	Males.	Females.
Brought forward from p. 256	92,038	30,094	85,960	28,358
18. Baroda,.....................	450	174	397	172
19. Mysore,.....................	340	193	536	266
20. Rájputána,.................	0	0	1,314	394
21. Cent. Ind. Regs., etc.,	7	6	59	21
22. Bombay States,..........	1,681	606	1,907	641
23. Cawnpur State,..........	0	0	4	2
24. Cochin & Perderkottal,	143	92	313	138
25. Travancore,...............	0	0	684	284
26. Central Prov. States,...	0	0	799	460
27. Bengal States,...........	1,799	750	1,471	577
28. N.-W. Prov. States,....	339	94	312	67
29. Punjab States,...........	2,241	613	1,462	458
TOTAL,............	99,038	32,622	95,218	31,838
	131,660		127,056	

The general reader may perhaps be assisted in understanding the foregoing Tables by the following Analysis.

PROVINCE.	MALES.		FEMALES.	
	Increase.	Decrease.	Increase.	Decrease.
1. Ajmír,.......................	......	3	1	......
2. Assam,.....................	2,719	......	693	......
3. Bengal,.....................	......	7,527	......	2,461
4. Berar,.......................	......	85	847	......
5. Bombay,...................	299	......	......	140
6. Sindh,.......................	......	41	......	27
7. Aden,.......................	1	......	......	......
8. Upper Burmah,..........	2,262	......	1,242	......
9. Lower Burmah,..........	272	......	99	......
10. Central Provinces,......	......	855	......	233
11. Coorg,......................	......	13	......	7
12. Madras and Small Feudatories,...........	......	874	......	664
	5,553	9,398	2,882	3,532

ANALYSIS OF STATISTICS.

PROVINCE.	MALES.		FEMALES.	
	Increase.	Decrease.	Increase.	Decrease.
Brought forward from p. 257	5,553	9,398	2,882	3,532
13. N. O. Prov. and Oudh,	......	339	......	412
14. Punjaub,	......	2,011	......	518
15. Quettah,	2	......	......	......
16. Andaman,	......	29	......	......
17. Hyderabad,	144	......	......	156
18. Baroda,	......	53	......	2
19. Mysore,	196	......	73	......
20. Rájputána,	1,314	......	394	......
21. Cent. Ind. Regs., etc.,	52	......	15	......
22. Bombay States,	226	......	35	......
23. Cawnpur State,	4	......	2	......
24. Cochin and Pardukta,	170	......	46	......
25. Travancore,	684	......	284	......
26. Central Prov. States,	799	......	460	......
27. Bengal States,	......	328	......	173
28. N.-W. Prov. States,	......	27	......	27
29. Punjab States,	......	779	......	155
	9,144	12,964	4,191	4,975
Deduct increase,		9,144		4,191
Net decrease,		3,820		784
	1891	95,218	1891	31,838
	1881	99,038	1881	32,622
Places,	15	14	13	13
Add increase,		15		13
Unchanged,[1]				3
	Total...	29	Total...	29

[1] Nos. 7, 15, and 16.

From the foregoing analysis, it appears that the number of male lepers has during the decade increased in fifteen places and decreased in fourteen. The number of female lepers has increased in thirteen places and decreased in thirteen : while in three places they are *in statu quo.* In seven places no returns are given for 1881.

The total number of lepers returned are as follows:— In the census for 1881, 131,660; in the census for 1891, 127,056.

It will be remembered that at a meeting held at Marlborough House on the 17th of June, 1889, His Royal Highness the Prince of Wales declared that there were 250,000 lepers in India, an estimate nearly double that which is indicated by the figures derived from the censuses of 1881 and 1891. That the estimate given by His Royal Highness is far nearer the mark than are the figures derived from the census, will appear probable from the following facts and considerations.

On the perusal of the census forms issued by the Indian Government, we find in Rule 14 that "white leprosy" is to be excluded from the infirmities to be returned by the enumerators. That rule is as follows:—
" If any person be blind of both eyes, or deaf and dumb from birth, or insane, or suffering from corrosive leprosy, enter the name of the infirmity in this column. Do not enter those blind of one eye only, or who have become deaf and dumb after birth, or who are suffering from white leprosy only."

Mr. J. A. Baines, Census Commissioner, writing 21st June, 1892, says that in the recent census, " the instructions were clearer (as to leucoderma, or white leprosy), and the exclusion far more strict."

The census form also contains the following direction:

"You are to make all the entries as the person himself, or his guardian, states, and not to dispute his statement."

Coupling this direction with the fact that now, for the first time within thirty years, the report has been widely circulated, that all lepers at large were to be segregated, that is, to be separated from their friends and from all they hold dear, we should infer that a large number of cases of leprosy would be suppressed, and not returned to the census enumerators, who are much more in sympathy with the afflicted members of their own race than with their official chiefs, and would be slow to aid in the perpetual incarceration of their friends; and among the respectable classes no one will admit that he is a leper, as an admission of this kind would involve loss of caste and social ostracism.

Moreover, leprosy is an insidious disease, and in its early stages cannot be diagnosed and detected save by experienced medical practitioners accustomed to treat this particular malady. Of the enumerators, not one in a hundred could detect a case of leprosy if he saw it, except when presented in its most aggravated and repulsive form.

Mr. H. A. Ackworth, Municipal Commissioner, Bombay, in a communication to me, dated 29th July, 1891, says:—"I have plenty of lepers in my hospital here who could not be identified as such unless they were completely stripped and examined by a trained eye." And a correspondent of the Calcutta *Daily News*, October 20th, 1891, writes:—"To my personal knowledge there are at least twenty-three lepers in this town who are not entered as such in the census papers. And it is quite probable that there are many more who have not been numbered."

A MISLEADING LEPER CENSUS. 261

A medical correspondent connected with the Army Medical Department writes to me, January 27th, 1892, "that the census form used in the census of 1891 is an enigma that the enumerators could not properly explain; and the remuneration offered was so small that it failed to tempt any of the better educated classes to volunteer for the census work. Consequently the papers were handed over to men who, for the most part, were too devoid of understanding to enable them to ascertain facts or comprehend the nature of the information required for filling in the returns." My correspondent adds that the wide-spread belief in India, that leprosy is a disease of venereal origin, induces all but the lowest classes to carefully guard against it becoming known, either to officials or to others, that any member of the family is a leper, and "even the threat of prosecution would not frighten them into publishing their terrible secret for census information." For these and other reasons there can be no doubt that a considerable portion of the leper population in India has been omitted from these returns.

Without endorsing the accuracy of the leper census in India, which is clearly misleading, it may be incidentally remarked that the proportion of vaccinations to population in India, as a whole, is less than in any of our crown colonies and dependencies, but is increasing every year, and, unless arrested, will soon produce the calamitous consequences exhibited in tropical countries where vaccination is general.[1]

---

[1] The births in India in 1890-1 were officially estimated at 40 per cent. of the population, or 8,244,101, and the number of children under one year returned as vaccinated in 1890-1 was 2,268,922, being about 27½ per cent. on the total number of births.

18

THE LEPER CENSUS IN THE LEEWARD ISLANDS, WEST INDIES, 1891.

A census has recently been taken for the Leeward Islands, which also minimises the number of lepers in a similar fashion. Referring to this, the St. Kitts *Lazaretto* of February 22nd, 1892, says:—

"The census return for the Colony is at length out, and a copy may be seen at the Public Library. It is openly said all over the Colony that this particular census is utterly unreliable; indeed, Mr. Fred Evans, the Colonial Secretary, says as much of the returns from Dominica. A glance at the tables which pretend to show the number of lepers in the Colony satisfies us that, in this particular, the returns are worthless.

"Some time ago, the doctors here were called upon to send in returns of the lepers in their districts. They did so, and sent in a list of 51 lepers. That this list was necessarily incomplete we fully demonstrated at the time. Since then, we have heard of several other lepers of whose existence no one here knew, and a few have been committed to the Leper Asylum. But in the face of the reports from the doctors, the compiler of these precious returns sets down the number of lepers at large in St. Kitts as thirteen and in Nevis as five!! How could Mr. Fred Evans or his clerk have got these figures? The census forms that had to be filled in by each householder contained no space in which the number of lepers could be entered. Therefore, we are forced to the conclusion that the opportunity of misrepresenting the leprous condition of this Colony was seized by those who were on the look-out for such a chance, the doctors' reports disregarded, and that some one invented these figures. . . .

"We guarantee to produce fully 50 (and perhaps 60) lepers who are now at large in this little island (St. Kitts).

"There are considerably more than 13 in Basseterre alone. As regards Nevis, the inaccuracy of the Colonial Secretary's figures is equally striking.

A MISLEADING LEPER CENSUS.

"We can give off-hand the names of a half-dozen lepers living there, and we have heard of many more, of whose existence we are assured by gentlemen who are above lying, and who have no motive for suppressing the truth.

"The figures for the other islands are also palpably incorrect and misleading. In Antigua 34 lepers are said to be under restraint. This is not true as regards the restraint, for according to Dr. Freeland they cannot legally be restrained, and as a fact they go and come from the so-called Lazaretto as they please. Only last month we saw several of them promenading the streets of St. John's. Then to set down 11 lepers only as being at large is a gross mis-statement, as every one in Antigua knows. We were up there quite recently, and satisfied ourselves that there are probably in all over a hundred lepers in that island.

"We accept the return of 3 lepers to about 500 people in Barbuda, but we confidently challenge the assertion that there are only 9 lepers in Montserrat and 3 in the Virgin Islands. . . .

"Anguilla, the Government says, contains 8 lepers. Will Mr. Fred Evans be surprised to hear that he has made the trifling error of only eighty per cent? We happen to know that if he had guessed 40 (for the whole thing is evidently pure guess work) he would have been pretty near the mark.

"The total for the Leeward Islands is, according to this eccentric statistician, 172. That for St. Kitts and Nevis is 98. We declare, speaking with a knowledge of the subject which is infinite compared to that possessed by Mr. Fred Evans, that there are at least 200 lepers in St. Kitts and Nevis; and that Sir William Haynes Smith knows. Taking St. Kitts, Nevis, and Anguilla, we would not be surprised if 250 free lepers could be ferretted out. Of Dominica and Montserrat we cannot speak so confidently, but we have interviewed many gentlemen from them, and the sum of our inquiries is that no one knows how many lepers there are in any one of them. It has been no one's business to find out, and therefore no one has given the least attention to the subject."

CHAPTER XI.

LEPROSY AND THE ABORIGINAL RACES.

IN 1889, during a visit up the river Essequebo, in British Guiana, the British Commissioner and resident Magistrate, Mr. Michael M'Turk, of Kalacoon, informed me that he had not the slightest doubt that leprosy was disseminated with the vaccine virus. He was intimately acquainted with a healthy family, in which one of the children was affected with leprosy by means of lymph taken from a child afterwards proved to be tainted with leprosy. The unfortunate victim of the state-enforced operation was isolated in a small building at the end of the garden at the parents' house, and ultimately succumbed to the disease. As an explorer, Mr. M'Turk has been much among the Indians, all his servants and boatman belonging to that race, and he had never known or heard of a case of leprosy amongst them. The truth of this statement is confirmed by a communication to me from Mr. Herman Klein, Acting Assistant Medical Officer, H. M. P. S., Potosi, British Guiana, dated June, 1891—" I have been about eleven years up the Essequebo river, and have never seen an Indian afflicted with leprosy." I received similar testimony at Bartica Grove from Mr. John Bracey, an Indian trader of twenty-nine years' experience among the Macousi and Wapisiana tribes. Dr. John

D. Hillis, F.R.C.S., formerly the Superintendent General of the Leper Asylum, Mahaica, in his work, entitled "Leprosy in British Guiana" (1881), says, p. 148 :— "With regard to this country one important fact is the immunity from leprosy enjoyed by the aboriginal tribes of British Guiana." This immunity from the disease is attributed to the circumstance that no Indian will allow himself or his children to be vaccinated. Dr. T. C. Taché, Titulary Professor to the Laval University, writing from Ottawa, Canada, in reply to questions snbmitted by the Hawaiian Government, June, 1885, says:— "There never was any case of leprosy among the Indians, although one of their principal villages is located in the endemic section, being contiguous to the parish of Nigavrick." Professor T. C. White says that in Tracadie, New Brunswick, more than 100 lepers were received at the hospital between 1849 and 1882 ; nearly all the cases were of French descent, and no Indian had fallen a victim.

Dr. J. E. Graham, in a report from the Government of Hawaii as to leprosy in New Brunswick (1886, pp. 114 and 140), observes :—" That the Indians have been the only race, of those inhabiting these localities in any number, which have remained so far exempt from leprosy. . . . The places in which the Indians dwell bear precisely the same character as those inhabited by their neighbours among whom the ailment has exercised its ravages."

Dr. Miguel Valladores, physician to the Lazaretto, Guatemala, says, in his report to the Hawaiian Government, 1886, p. 174, "that it is almost an unheard of thing for an Indian to be afflicted with leprosy."

Drs. Vlagthoes and Mayrinck state that, previous to the discovery of Brazil, leprosy was unknown among the Indians.

Dr. Alzevedo Lima, in a letter on "The Leper Hospital of Rio de Janeiro," dated June 1st, 1891, Rio de Janeiro, says:—"On consulting all the documents and books written by travellers and missionaries in Brazil, we find no mention, either direct or indirect, of any prevalent complaint among the Indians which might be attributed to leprosy. Even nowadays those who live a savage life away from all contact with civilised society are not attacked by this disease, while those who have left their woods for peopled centres, together with their descendants, are, according to observations, occasional victims."— *Journal of the Leprosy Investigation Committee, December, 1891, p. 22.*

In a communication to me, dated Rio de Janeiro, May 20th, 1892, Dr. Lima says:—"Now, about the Indian races, those who live away altogether, without any interference or intercourse with civilisation, their freedom from leprosy can be explained not only by the absence of the Jennerian vaccination, but also by the non-intercourse with people capable of being the conductors of the germs of the disease." I have personally met with races of Indians in South America, amongst whom, though living amongst lepers up to this date, no cases of leprosy have occurred. This immunity is attributed by old residents, one a physician, to the circumstance that they will not allow themselves or their families to be vaccinated.

Dr. J. Z. Currie, secretary of the Provincial Board of Health, Fredericton, New Brunswick, in reply to a communication from me, dated January 2, 1892, as to the vaccination of Indians in New Brunswick, Canada, says :—" There has been no outbreak of small-pox among the Indians for some time. However, in almost all instances they object to vaccination. Four cases of small-pox occurred in this Province during the past year among white people."

The main object of this evidence is to show that in countries where leprosy is endemic, the indian tribes who reject vaccination escape the plague. In New Brunswick it would appear that they also escaped small-pox, while vaccinated white persons have been attacked with the disease.

CHAPTER XII.

VACCINAL DISEASES IN SOUTH AFRICA.

VACCINATION was made compulsory in Cape Colony by Act of the Legislature in 1882. Very soon the deleterious effects of the virus were exhibited. The London *Daily News*, March 5, 1884, says that owing to impure lymph there had been many cases of illness from vaccination; but, later on, the natives were vaccinated on an extensive scale. The Public Health Act, No. 4, Cape of Good Hope, dated September 6, 1883, contains the provisions of a vaccination law of a stringent and despotic character. Section 60 states— " No person who has not been vaccinated shall be appointed, or, if appointed prior to the taking effect of this Act, promoted to any office in the public service." Section 61 provides that " Every child, admitted to any school which shall be maintained or aided by any grant from the public funds, shall be vaccinated by the District Surgeon or by a vaccinator specially appointed, unless such child shall have been previously vaccinated." The penalty for non-vaccination is £2. The law contains other oppressive clauses, thus incorporating in one act the worst features of the English, American, and Continental vaccination enactments.

All this time, while the vaccine poison was being forced into the blood of the defenceless natives, laying the foundation for the disorders which speedily followed, nothing was said or done to remove the causes which developed the outbreaks of small-pox, the fearfully insanitary condition of the town in which the pestilence abounded, the fruit of long-continued filth and neglect, scarcity of water, foul, unkempt streets, seas of mud in the winter and hurricanes of dust in summer, and, worst of all, a population ignorant of the commonest instincts of decency.

The *Cape Times* reported that Cape Town was buying its experience at a heavy price. Within a short time of the introduction of compulsory vaccination, spreading with accelerated industry this tainted virus distilled from the bodies of a filthy population, we read of the spread of leprosy, and of the alarm created in the Colony among those who had observed its destructive progress. A not uncommon experience is to hear of cases of leprosy in families where there is no taint of the disease, and where the afflicted member has never come into contact with lepers. The late American Consul, Mr. James W. Siler, of Cape Town, in his official report to his Government, No. 79, June, 1887, records a case of this description; and, as vaccination is obligatory, the obvious causation is that the disease has been transferred in the vaccine virus. Mr. Siler says:—" A case with which I am well acquainted will illustrate its seemingly mysterious power of propagation. In one of the oldest and wealthiest Dutch families in this Colony the mother is a confirmed leper, of the type described as 'tubercular' by Dr. Atherstone, before alluded to. The father and a large family of strong, healthy, grown-up sons and daughters show not the slightest taint. I have several times enjoyed the hospitality of this family, and availed myself of the opportunity thus afforded of inquiring into this melancholy case, with the view of a possible solution. I am assured that neither on the side of the father nor mother a case of leprosy had ever occurred in their families, and they are able to trace their genealogy back at least one hundred years."[1]

The conclusions of the Select Committee regarding the increase of leprosy in South Africa[2] derive confirmation from the individual reports of missionaries, clergy, travellers, and district surgeons in South Africa, and this increase is specially observable where vaccination has been extensively practised. Referring to the reports of district surgeons, published at the Colonial Office, Cape Town, and presented to both Houses of Parliament, I find the following relating to leprosy and to syphilis, a disease, according to various authorities, pathologically allied to leprosy :—

The medical officer for *Herbert* (Report, 1885), says :— "During the year, small-pox, syphilis of a particular type, and leprosy, have been the prevailing epidemics. The two last named are still prevailing to an alarming extent." It would appear that vaccination, as usual, was resorted to on account of the small-pox, but the medical officer reports that "the difficulties in carrying out arm-to-arm vaccination seem insurmountable."

---

[1] The *British Medical Journal*, July 5, 1890, under the head of "Reports — Liverpool Workhouse Hospital," communicated by Dr. Cunningham, Senior Medical Officer, gives particulars, with copy of photograph, of an "interesting case." C —— L ——, aged 46, who has a husband and six children, all of whom are healthy, and, until she became affected with leprosy, had never suffered from any disease. "*She has never known or seen anybody with the same* or similar disease." The italics are mine.

[2] Chap. I., pp. 68 and 69 of this report.

Cala.—The District Surgeon (Report, 1885), says:—
"About two years since, there were many cases of swollen arms, and some deaths after vaccination," and observes: "There is also the danger of inoculating syphilis."

Aliwal North (Report, 1885).—"Vaccination has been extensively performed amongst both Europeans and natives." In the following year (Report, 1886), the same officer remarks that "small-pox has raged," of which he has treated about 450 cases, and adds that "syphilis has made vast strides." Two years later (Report, 1888), we read :—"Syphilis is still very prevalent. I have frequently drawn the earnest attention of the Government to the sad havoc this disease is dealing amongst the inhabitants."

While showing the utter failure of the extensive vaccination practised in 1884 to prevent the serious epidemic of small-pox which occurred the following year, these three official reports show how the most loathsome of diseases are disseminated by the vaccinator's lancet. The District Surgeon urges the Government to appoint a Commission of Inquiry.

Alexandria (Report, 1887).—"Leprosy is certainly spreading rapidly." In 1890, the District Surgeon reports that the state of leprosy demands urgent attention.

Port Elizabeth.—The medical officer says (1887) :—
"There is a growing aversion to it (vaccination), partly due to an underlying current of belief in the possibility of obnoxious disease being propagated by it." The Report for 1891 states that during the year the presence of leprosy was gone into, and six cases reported to the Government.

Caledon.—The District Surgeon reports (1888) that he has vaccinated close upon 800. There are about twelve cases of leprosy in the district, and it appears that syphilis is so increasingly prevalent that an hospital is needed for syphilitic patients.

Cape-Wynberg.—(Reports, 1887-8). The medical officer says:—"Of the great increase of leprosy there can be no doubt; it is obvious to the casual observer that Europeans as well as natives are afflicted with it. . . . The number of cases vaccinated has been about 100." The following year (1889), the District Surgeon says:—" Leprosy is becoming far more frequent in the neighbourhood " . . . and adds:—"The question of leprosy is one of the most serious the Government have to deal with. . . . Vaccination has been thoroughly carried out throughout the district."

Malmesbury (Report, 1888).—"Leprosy is slowly but surely gaining ground." This officer reports that he does not think it advisable to vaccinate in the district.

Cradock (Report, 1887).—"Fifteen cases of leprosy have occurred, all in an early stage."

Paarl (Report, 1887).—"Leprosy is on the increase." Report, 1890.—"Leprosy is spreading."

Glen Grey (Report, 1889).—The District Surgeon says: —" I have incidentally about a dozen cases of leprosy, some of these of quite recent origin." . . . "A centrally-situated leper hospital is imperatively required."

Kokstad (Report, 1886).—"Vaccination during the past year has been in several districts well carried out." Two years later (Report, 1888).—The Medical Officer for this district writes as follows:—"Leprosy is still very much on the increase. There are at least fifty cases in Kokstad itself." . . . "It is deplorable to see these wretched victims dependent on the public charity for a bite, whilst the Government will do nothing for their alleviation."

Stellenbosch.—The District Surgeon (Report, 1889,) remarks that he has just vaccinated two hundred children. He reports 20 cases of syphilis and six cases of leprosy. In 1890 the same officer returns nine cases of leprosy, and adds "but no doubt there are a few more unknown to me."

Stockenstrom.—Referring to leprosy (Report, 1890,) "I have seen persons without hands paying their quit rents, holding the money on the stumps of their arms."

Somerset East (Report, 1890).—"We have a good few lepers here, as already reported."

One experienced district surgeon told me that he had, again and again, year after year, called the attention of the Board of Health to proofs of this terrible havoc wrought by arm-to-arm vaccination, and had advocated its suppression in the interests of public health. A careful examination of the official documents would show that the facts incriminating vaccination have not been allowed to appear.

When making inquiries regarding etiology and spread of leprosy in South Africa, I was generally referred to the Rev. Canon Baker, of Kalk Bay, Cape Colony, as a high authority on the subject, and one who had probably devoted more attention to it than any other resident in the Colony. Canon Baker had in 1883 given evidence before the Select Committee of the House of Assembly, Cape Town, and presented a statement of his views, which appeared in Appendix A, pp. 1-9. Since then he has continued his investigations and accumulated a considerable body of facts bearing on the subject. Vaccination, he says, is carried out in the Colonies in a most careless and perfunctory manner. He has seen the operator pass his lancet from one arm to another without the smallest attempt to disinfect the instrument or discriminate between the diseased and the healthy, in districts where both leprosy and syphilis are endemic. From other reliable sources I am satisfied that this is the rule rather than the exception. Canon Baker believes that leprosy is chiefly communicated by means of inoculation, and that arm-to-arm vaccination is a prolific cause of the spread of this fearful plague in South Africa.

The Colony of Natal passed Vaccination Law No. 3 in 1882, and Law No. 10 in 1885. Penalties for non-vaccination £5. In a communication from Archdeacon Colley, dated Natal, August 25, 1885, I learn that hundreds of summonses were issued in vain upon the colonists, but the natives were vaccinated by thousands; one operator would get through two hundred a day.

While the vaccination laws for several years have not been enforced against the white population in Natal, all the natives are vaccinated either under persuasion or threats, the operation being carried out in the usual careless manner, with arm-to-arm virus taken from native children without previous examination, and not the slightest attempt is made to clean or disinfect the lancets after each operation. Hundreds of natives, as I am informed on unimpeachable authority, have died of blood-poisoning and of inoculated diseases.

A member of the Legislative Council, Sir John Bisset, reported in Parliament that many were " blood poisoned, presenting a horrible sight, and dying masses of corruption." In January, 1891, leprosy disseminated in this way was discovered in fifty kraals in one electoral division alone. The natives in their simplicity submit to vaccination, being told that it was the "Incosi" (King) that ordered it, and this was the way the white man secured himself against the plague of small-pox.

As the Government of Natal does not publish reports from the District Surgeons, and appears to be indifferent as to the suffering and mischief caused by the vaccinators, I found it difficult to obtain further details.

CHAPTER XIII.

A VISIT TO THE LAZARETTO, ROBBEN ISLAND, SOUTH AFRICA.

ON the 9th of February, 1892, after obtaining a permit from the Colonial Office, I took passage in the small but stoutly-built little tri-weekly steamer "Tiger," from Dock Basin, Cape Town, for Robben Island. Amongst our fellow-passengers were visitors, merchants, a clergyman, a singing lunatic in the custody of a warder, and officials connected with the island. Our cargo consisted of fruit, poultry, beef, and other stores for a population of about 700 persons. In less than an hour after starting, we cast anchor opposite the island; and, there being no landing-stage or jetty, we found small-boats, managed by convicts clad and numbered in penal costume, awaiting our arrival. These singular-looking boatmen rowed us near the shore, and then carried the male passengers on their backs and the women in chairs through the surf on to *terra firma*. The convicts on the island, about one hundred in number, are said to be chiefly murderers and diamond stealers. There is a strong current between the island and the mainland. Only one attempt has ever been made by a convict to escape by swimming, and the attempt cost him his life. The first building I entered for the purpose of making inquiries proved to be the female ward of the lunatic asylum. Here I was referred by the attendant to the office of the medical superintendant for another permit, which was granted. Dr. S. P. Impey has medical charge of the convicts, lunatics, lepers, and attendants, comprising the entire population of the island. I found him busily occupied with one caller after another, examining and signing papers from different departments; and, instead of engaging his attention at the moment, I expressed a hope that, as I was interested in what is popularly known as the leprosy question, he would be able to see me later in the day. He readily agreed to see me for this purpose at two o'clock. On my way to the leper wards I looked in at the little church, a rather pretty edifice, adorned with scripture mottoes disposed in large letters around the galleries, where the Rev. W. W. Watkins, the successor to the late Mr. Wiltshire, ministers to his singular congregation. I then proceeded to the male leper wards, meeting with convicts and lunatics on my way. The sun was scorchingly hot, but there was a good breeze blowing, and the ozonised air from the ocean was gratefully invigorating. I should think that the island, although only a sandbank of about 1200 acres in extent and almost devoid of vegetation during the hot season, would be very salubrious if the conditions for health were observed. The water supply is excellent. The first lazaretto building I entered—one of the old wards devoted to male lepers—was by no means an inviting structure. It was a large shed with low-studded walls containing a double row of beds, upon which the lepers were reclining in every variety of posture. The air was close and noisome with the evil effluvia of decaying living bodies—death in life—supplemented by the odour of influenza, which at this time was raging throughout the island. Of one hundred male lepers, no fewer than eighty, on the day of my visit, were down with this distressing disease. It has been said that a leper hospital, with its handless, footless, ulcerated, feature-swollen, and distorted patients, is as horrible a sight as a field of battle; and to the misery inseparable from a repulsive and incurable disease were now added the effects of a depressing epidemic—a most heart-breaking spectacle surely. Investigation as to the causation of leprosy under such circumstances was not a promising outlook. The majority of the inmates belong to the poorer classes of the native or mixed races, and are unable to speak English. In some cases the destructive disease had invaded the larynx, and they could only converse in a whisper. It was as painful to hear as to look at them. I must add that a new and much more spacious and suitable hospital, substantially built of stone, has been erected and is now used for the accommodation of fifty patients. Here I found a better state of things. Another building of like character is in course of erection; so it is evident that the authorities, while apathetic as to the causation of leprosy, have begun to realise the importance of doing what they can to render the condition of these helpless patients as tolerable as possible. I shall make no attempt to describe the state of the sufferers. I have seen more repulsive cases in lazarettos in other countries, and have seen them both under more favourable and under less favourable conditions. I am of opinion that the lepers and lunatics should be removed to separate quarters, and this barren island should be used exclusively for convicts. Leprosy and lunacy are not criminal, though the former is often caused by the criminal conduct of those responsible for vaccination and for insanitary neglect, and the latter by the temptations of the dram shops, where the vilest and most health-destroying liquors are regularly dispensed. The sufferings of these unfortunate people are sufficiently severe without their being compelled to associate and spend their unhappy lives with convicts. Notwithstanding these difficulties, I was able to hold a brief converse with about fourteen of the inmates who were well enough, amidst their complicated maladies, to understand and reply to a few simple questions. One of the three leper cooks of the establishment kindly acted as my conductor, and pointed out to me those who could speak English. Several were at work as shoemakers and tailors, but the rest had no employment—and, I was informed, did not want any— to relieve the terrible monotony of their painful existence. Some of the inmates were suffering acutely, and needed tender nursing and such devotion as the Moravian Brothers and Dominican Sisters bestow upon lepers in other hospitals. Several of the patients make pets of a harmless snake, a variety of python common to the island, which they keep in cages near their beds. After I had been introduced by my conductor, and had exchanged salutations with a few words of inquiry, my interrogatories were as follows:—Where were you born? How long have you been a leper? When did you come to Robben Island? Have you had the smallpox? Have you been vaccinated? How long after vaccination did the leprosy appear? The cook, Christian Choutsee, a native of Cape Town, said he had been a leper four years, and the leprosy broke out "two years after the doctor stuck me in the arms." The answers to the last question were "two years," "three years," "a few years," "two or three years," "after the second vaccination." "Was vaccinated in 1879; leprosy appeared in 1883." "Vaccinated three times, last vaccination during small-pox epidemic in 1878; leprosy attacked me in 1887." "Vaccinated twice, first when twenty years of age. Leprosy appeared between first and second vaccination." "Vaccinated during small-pox epidemic of 1878. Leprosy broke out on me about a year after." "Vaccinated when a boy of between eight and nine years of age; have been a leper fourteen years; present age, twenty-six." After going through the several male wards, I took an opportunity of making calls upon several persons, including the clergyman, the superintendent of the male wards, and other officials. No one seemed to doubt that leprosy was spread by vaccination. The superintendent of the dispensary, who requested me not to publish his name, gave me particulars of the case of Augustus Lewis, of Cape Colony, who died of leprosy at Robben Island, the disease having been induced by vaccination. I had now been pursuing my inquiries several hours, and as the female leper ward was about a mile distant, and approached only by a rough, stony track, I was obliged, in order to keep my appointment with Dr. Impey, to forego my intended inspection of this department. Dr. Impey is deeply interested in the pathological side of the leprosy question, and his position as superintendent of the largest leper institution in Africa affords him ample opportunity of pursuing his investigations. He has practised twelve years in South Africa as a District Surgeon and Physician, and apart from his clinical experience gained through observations, and the medical care of the population of the island, he has found time to study the literature of the subject. He regards Dr. John D. Hillis's "Leprosy in British Guiana" as the most valuable and important work he has read. I called his attention to the cases of invaccinated leprosy cited by Dr. Hillis in the volume referred to, and he expressed no surprise at this, having come to a similar conclusion through his own personal researches in different parts of the Colony. Dr. Impey informed me that after careful investigation he had clearly traced to vaccination four out of twenty-eight cases of leprosy, which he had examined in the female ward. One of these still shows leprous discolourations at the point of vaccine inoculation, the disease having exhibited itself two years after vaccination. Dr. Impey will continue his investigations as to the causation of leprosy amongst the remainder of the patients; a procedure, let me observe, almost unknown at similar institutions. It is needless to say that the report of these investigations will be awaited with much interest. Although a believer in the protective value of vaccination as a mitigator of small-pox, Dr. Impey has met with so many cases of invaccinated syphilis and leprosy that he has felt it his duty in his reports, extending over a period of eleven years, to point out the mischief already perpetrated by this mistaken procedure; and he has called upon the Government to legislate for the immediate and total suppression of arm-to-arm vaccination. Dr. Impey considers that leprosy is contagious by actual inoculation, cases of which had occurred at Robben Island, and that to a certain limited extent it is hereditary.

CHAPTER XIV.

THE SEGREGATION OF LEPERS.

So far, I have said nothing concerning the growing demand for compulsory segregation of lepers. It is admitted on all sides that the forcible deportation and confinement of well-to-do lepers would be impracticable, and already there are too many laws which are cruel and oppressive to the poor, but which, by the wealthy, are easily evaded. So far as the well-to-do are concerned, the law of enforcing segregation of lepers at Molokai, as I have already shown, is an admitted failure, and the act of separation of the poor from their friends is the most heart-rending and painful experience which, in a tolerably long life, I have ever witnessed.

In the "Report on Leprosy by the Royal College of Physicians, 1862," I find the following:—" The Committee, having carefully considered the replies already received, are of the opinion that the weight and value of the evidence they furnish is very greatly in favour of the non-contagiousness of leprosy. The Committee can only repeat the statement made in their former report to the College, that the replies already received contained no evidence which, in their opinion, justified any measure for the compulsory segregation of lepers." Acting on this opinion, the Duke of Newcastle issued a circular to the Governors of the Colonies, stating "that any laws affecting the personal liberty of lepers ought to be repealed, and any action of the Executive Government in enforcement of them, which is merely authorised and not enjoined by the law, ought to cease."

Dr. George L. Fitch, formerly Medical Superintendent, Leper Settlement, Molokai, Hawaii, says:—"Segregation began in 1866 in Hawaii, and since that time has been followed out with a really brutal severity. At no time since the inauguration of the system has the proportion of cases segregated fallen as low as one-half, so far as I could find out, and I had the fullest opportunity to know of any one. The white population there are terribly in earnest, and, as they control the policy of the country, they have exercised every ingenuity in this matter, so that it may be considered certain that an average of two-thirds, at least, have been for the entire period under a restraint much more pronounced than is the case in Norway. Yet there is not the slightest evidence that the disease has decreased; at least, I know of no such evidence. From all I hear from there, the proportion of lepers continues as great, if not greater, than it has been for years. That the disease does not manifest as severe symptoms as formerly is certain, but I know of no reason to believe that the percentage of those afflicted has lessened at all. On the contrary, both the total number, and the proportionate number of cases, would seem to have steadily increased. . . . Of late years, so bitter a feeling has grown up among them in opposition to segregation, that in quite a number of instances the lepers and their friends have risen in arms to resist the officers sent to apprehend them. . . . Bring these three facts together. In India, up to 1815, lepers were buried alive to get rid of them; and still the disease persists. In Norway the disease is disappearing without segregation, for putting two cases out of five into hospitals, where they are allowed to carry on their handicrafts, and selling the products to those outside, cannot be called segregation. In Hawaii, where as thorough segregation as the Government, aided by public opinion, can enforce, is carried out, the disease steadily increases." — *New York Medical Record, September 10, 1892. Art., " Etiology of Leprosy," p. 301.*

Dr. George Thin, in his recent work on " Leprosy," though an advocate for the seclusion of lepers, shows himself alive to the difficulties of compulsion. In pp. 257-8 he says: — "A law enforcing the compulsory isolation of lepers can only be effective in any country where leprosy is common, if it is strongly supported by public opinion. Those who have little practical experience of lepers and leprosy must not forget that for a considerable time, and often for years, the stricken member of a family suffers comparatively little, requires little attention, is not specially repulsive in appearance, is as full of love for his parents and brothers and sisters, and in return is as much loved by them, as if he were not afflicted by the disease. To realise what compulsory isolation in an asylum of all the lepers in a country would mean, when such cases are considered, it is only necessary to apply in imagination the same law to consumptives when their disease runs a slow insidious course. What would the consequences be in England if a law were passed that every husband, or wife, or child, who developed a slight cough, attended with weakness, and in whom slight physical changes were detected in the apex of one lung, should, on the strength of what his probable fate would be several years afterwards, be immediately and forcibly conveyed from his family with no, or scarcely any, hope of ever again rejoining them? Imagine the evasion, concealment, and subterfuge that would be practised, and the difficulty, if not impossibility, of passing a law which would be effective! As a matter of fact, already, and with no compulsory isolation, in all but the very poor, leprosy is in most countries concealed as long as it is possible."

A medical practitioner who has resided several years at Honolulu and Molokai informed me that he personally knew of a number of well-to-do lepers, some occupying prominent positions, including several Europeans, who from political and other influences with officials of the Government were allowed to be at large, and it was not intended to disturb them. Nor is it considered possible to amend this partial method of dealing with the difficulty, especially as the natives do not believe in contagion. At the Leper Asylum in Ceylon, as also in the West Indies, I found that only the poor were segregated in the lazarettos, and in every country I have visited, the compulsory segregation of *all* lepers is considered impracticable. Moreover, the experience in Hawaii has been the reverse of encouraging. Mr. R. W. Meyer, agent of the Board of Health at the Leper Settlement, Molokai, in his report dated April, 1886, observes that segregation has now been practised for twenty years, and the result is that there are as many lepers as ever ; more than at the commencement.

In an article in *The Lancet*, August 26, 1882, p. 318, commenting on a report on leprosy in Hawaii, and referring to the segregation of lepers, the writer says: "Nothing can, we think, call for action such as is described in certain parts of the report, and which has also called forth a protest by the Assistant Attorney-General of the kingdom. This gentleman describes how people supposed to have leprosy have been taken summarily from their houses by the police authorities, and have, without a moment's preparation, been ordered into boats, and conveyed across to one of the island settlements, where, as he says, they, 'are practically doomed to death.'"

That compulsory segregation cannot be carried out save by setting aside every humane feeling is admitted by those who are familiar with its operation. Thus, in his official report to the Board of Health, Honolulu, Mr. R. W. Meyer observes: "After the most careful consideration, I find that this is a question which involves a great principle, and which duty to oneself and his fellow-men alone should decide; a question which demands the absolute setting aside of every influence resulting from a feeling of sympathy with the unfortunate sufferers."

In the "Report of the Select Committee on the Spread of Leprosy in South Africa" I find the following:—

> Q. 52. "Do you believe there will be much difficulty in compulsorily removing these people?"— "Yes, there will be great difficulty. They will do what they can to conceal cases. They will steadily deny that any one in the house is affected by the disease. I have found the most undoubted and notorious

cases denied. They have a great aversion to remove. Most of them are married, and, in addition to the natural repugnance at parting from their wives, their sexual passions are particularly strong—in fact, they become, both mentally and physically, a lower type."—*Witness*, Dr. H. C. Wright, June 27th, 1889, District Surgeon at Wynberg.

Q. 92. "Take the case of a respectable man, educated and intelligent, would you separate him from his wife and family and remove him from his home?"—"I am afraid there is no help for it."—*Witness*, Dr. Simons, District Surgeon of Malmesbury, July 4th, 1889.

Q. 192. "I think that husband and wife should be separated, and that is a hard case, but necessary."—*Witness*, Dr. Beck, of Roudebosch, July 4th, 1889.

LEPROSY REPRESSION ACT.

Under this heading, the *Cape Times* of April 22, 1892, gives the following:—

ACT OF 1884 TO BE ENFORCED.

The Government have decided to promulgate the Leprosy Repression Act of 1884, and as at present decided the enforcement of the provisions of the Act will take place during the ensuing month. This action has doubtless been suggested by the disclosures of the census of a year ago, when the number of lepers in the Colony and the Transkei was shown to be 625—at least double the number casually reported from the various districts of the Colony. Since the taking of the census the number of known lepers has been increased to 664, and the promulgation of the Act will therefore necessitate the immediate provision of accommodation for the large number of sufferers who are still at large. The following will show the distribution of lepers in and out of hospital at the date of the census, in April, 1891 :—

IN HOSPITAL.

The Colony proper, as constituted in 1875 : European or white, 21 ;—14 males and 7 females; other than European, 97 ;—77 males and 20 females.

OUT OF HOSPITAL.

The Colony, as constituted in 1875 : Europeans, 30 ;—15 males and 15 females; other than Europeans, 256 ;—130 males and 126 females. Province of Griqualand West : Europeans, nil ; other than Europeans, 17 ;—males, 12 ; females, 5. Transkeian Territories : Europeans, nil ; other than Europeans, 204 ;—118 males and 86 females.

ROBBEN ISLAND.

A special report upon the lepers and accommodation for such upon Robben Island shows that, excluding the 54 coloured sufferers recently received from the Orange Free State, the number now on the Island is 162, of whom 16 are Europeans and 146 coloured persons. There are therefore over 500 lepers in the Colony, Griqualand West, and the Transkei still to be provided for. It is interesting in this connection to note that during the past year no fewer than 139 persons have voluntarily sought refuge within the leper wards of Robben Island Hospital. Of these 85 were males, 6 being European and 79 coloured persons. The European females admitted number 3 and the coloured females 51. During the same period 1 coloured male and 1 coloured female have been discharged from the island, whilst there have been 29 deaths, viz.: Five European males and 16 coloured males, and 8 females, 1 white and 7 coloured. In view of the demand for accommodation, which will arise upon the promulgation of the Leprosy Repression Act, the wards at Robben Island are being largely extended, and we are informed that by the end of May the leper hospital will be completed and provision made for all the lepers known to exist in the

Colony and the Province of Griqualand West. The Transkeian lepers, all of whom are coloured, will be centred at Engoobo, where a large asylum is now in course of construction.

The Leprosy Repression Act has since been promulgated in Cape Colony, and a systematic hunt for lepers has been carried on with the usual distressing concomitants—separation of parents from children, husbands from wives, friend from friend. A considerable number of lepers are in close concealment, carefully hidden by their friends. The well-to-do lepers have not been molested.

CHAPTER XV.

SELF-DEVOTION TO LEPERS.

HEROISM and self-sacrifice in the interest of humanity, like that displayed by the brave Father Damien, are, happily for the human race, by no means of unusual occurrence, as is shown by the devotion of the Dominican Sisters at the Leper Hospital, Port of Spain, Trinidad; that of the Franciscan Sisters, from Syracuse, United States, at Molokai, Hawaii; and the Sisters from Montreal, who tend the lepers at Tracadie, New Brunswick.

Dr. M'Laren, Nova Scotia, in a paper on " Leprosy in New Brunswick," read before the Medical Society of that Province, says :—" In 1868 a community of nuns from the Hotel Dieu, Montreal, most unselfishly took charge of the nursing of the sick, and the work is done faithfully and cheerfully under the Sister-Superior Mother, Saint Jean, and the lepers are much better attended to than formerly. The patients have plenty of freedom with grounds of eleven acres to garden, fish, etc."—*Maritime Medical News, Halifax, Nova Scotia, July, 1890.*

Of another hero, the *St. James's Gazette*, London, September 30, 1891, says :—" The last mail from Japan brings news of the death of Father Testevuide, a Japanese Father Damien. He was a member of one of the French congregations, and was sent to work in the Japan mission field. In 1886, during his labours in the interior, he came across a case of leprosy, which so aroused his feelings that he determined to give himself up to the task of ameliorating the condition of Japanese lepers. A woman of about thirty years of age, having developed leprosy, was abandoned by her husband, and, as the disease advanced rapidly, she was placed in solitude in a loft over a rice mill. In course of time the ravages of the disease rapidly increased, and she lost her sight. In this condition she was found by Father Testevuide, who was working in the district. He visited her constantly, and by reading and conversation sought to alleviate her misery; but he soon came to the conclusion that in her then condition she could receive but little relief unless she were placed in a hospital. There was no leper hospital in Japan, and the ordinary hospitals were naturally, for the most part, closed to such cases. From that time he devoted all his energies to the establishment and organisation of a leper hospital. Having succeeded in awakening public sympathy in the country, he collected sufficient money to build on the lower slopes of Mount Fujii a hospital which has for some years past been in full working order. His example was followed by some native philanthropists, and there are now three leper hospitals in the country. Father Testevuide's labours had undermined his health, which a visit to Hong Kong failed to restore, and he died there on the 3rd of August."

The Moravian Brothers have been sending missionaries of both sexes to live with and work among the lepers in the West Indies, in South Africa, and in Syria, devoting themselves sedulously but unostentatiously to this noble service during the past half-century. They were the pioneers in the effort to ameliorate the condition of these unfortunate sufferers. The Moravians have a leper asylum in Jerusalem, founded, managed, and largely supported by themselves.

The *Yorkshire Post* (Leeds), January 7, 1892, briefly refers to the death of an Anglo-Indian Father Damien reported from India. The victim was the Rev. W. D. Dalrymple, a Presbyterian missionary, who, having gone on a mission to lepers, contracted the disease some two years ago, and died last month at Rampur Beauleah, Bengal. Although his sufferings were indescribably great, he is said to have borne them with fortitude and resignation, and never once turned from the task he had set himself. Truly the age of martyrs is not past.

In the description of a visit to the Leper Hospital at Maracaibo, Venezuela, Consul Plumacher, of the American Legation, in a recent report to his Government, says:—"It was truly a sad sight to see deformed, mutilated trunks, with scarcely vestiges of extremities, seated before the camera; and there was something pathetic in the almost universal request to be supplied with pictures of themselves, which could only be constant reminders of their hopeless afflictions. In addition to the individual photographs, various large groups were taken, with an effect both sad and grotesque. There is one bright spot, however, in the dark picture of misery; this being the devotion and self-abnegation displayed by the near relatives of many of the sufferers, who, although enjoying themselves the blessings of health and strength, cheerfully submit to perpetual imprisonment, in order to minister to the wants of their husbands, mothers, and other relatives, thus alleviating their woes by their companionship and care. Many examples of this are seen to-day on the lazaretto island, and it speaks well for human affection that, even when the loved one has become a loathsome mass, conjugal ties and the claims of blood rise superior to the fear of contagion and the repulsive surroundings."

Miss Kate Marsden's labours among the Maories in New Zealand, and her extraordinary journeys through Russia, and among the wild tribes in remote parts of Siberia, with a view of learning the condition of the outcast lepers in districts where the disease is prevalent, are well known through the reports in the *Times* and *Pall Mall Gazette*, which have been extensively copied in English and colonial journals. Miss Marsden's object is to learn by personal observation the condition of the lepers; to discover, if possible, methods of mitigating their sufferings, and to collect funds for the establishment of leper hospitals. Some time ago, Miss Marsden consulted M. Pasteur to see whether inoculation as a cure of the disease might not be resorted to. M. Pasteur held out no hopes of amelioration in that direction, nor did he suggest any other. It does not appear that Miss Marsden has made any inquiries regarding the effect of vaccine inoculation in disseminating the scourge, although, in districts like Dorpat in the Baltic provinces, the lepers are reported to be rapidly increasing, and already form as large a portion of the population as 17 per thousand. The growth of the disease has been co-incident with the development of the Jennerian practice. Miss Marsden's knowledge of the Russian language, and her earnest desire to get at the root of the evil, would enable her to break through official apathy, so obstructive of truthful research, should she be induced to undertake such a mission. It is surely as laudable to arrest one admitted source of the mischief as to prosecute an almost hopeless search for remedies.

Another lady, Mrs. Alice Hayes, has also done much to direct public attention to the neglected condition of the lepers in India, particularly the Europeans and Eurasians of Calcutta, who hide themselves and their sufferings from the public in the large cities, and refuse to consort with native inmates of existing institutions.

CHAPTER XVI.

THE LEPROSY INVESTIGATION COMMITTEE.

By reason of the reports of the serious increase of leprosy in various countries, and the public interest excited by the self-sacrificing labours and death of Father Damien, an influential committee was convened for the purpose of investigating the causes of this recrudescence.

The first meeting was held on the 17th June, 1889, at Marlborough House, under the presidency of the Prince of Wales. On the 13th January, 1890, a subscription dinner was held at the Hotel Metropole, London, at which more than a hundred persons interested in the project sat down, and subscriptions amounting to over £2500 were announced.

This fund ultimately reached about £7000. The following resolutions were adopted at a meeting of the General Committee held at Marlborough House on the 30th June, 1889 :—

(1) That a sum of £500 be appropriated to a memorial to be erected to Father Damien in some public place in the Hawaiian Islands.

(2) (*a*) That a fund be formed, the interest of which shall be devoted to the medical treatment and care of indigent British lepers in the United Kingdom.

(*b*) And that a sum of money be set apart and placed under the control of trustees for the endowment of two studentships, one student to make the United Kingdom and the remainder of Europe his field of investigation, and the other to go abroad and study the disease in China, the Colonies, and elsewhere. The studentships to be held for a period of three years, to be renewed by the trustees if thought desirable.

(3) That a Commission be appointed, for not less than one year, consisting of three members, one to be named by the Royal College of Physicians, one by the Royal College of Surgeons, and one by the General Committee of the National Leprosy Fund, to go out to India for the purpose of investigating the disease of leprosy there, and that the Indian Auxiliary Committee be requested to add two members to this Commission.

The following Commissioners were appointed :—

Nominated by the Royal College of Physicians—
BEAVEN NEAVE RAKE, Esq., M.D. (London), M.R.C.S. (England), L.R.C.P. (London), Government Medical Officer and Medical Superintendent of the Trinidad Leper Asylum.

Nominated by the Executive Committee of the National Leprosy Fund—
GEORGE ALFRED BUCKMASTER, Esq., M.A., B.Ch., M.D. (Oxford), D.P.H. (Diploma of Public Health, Oxford), M.R.C.S., L.R.C.P. (London), formerly Radcliffe Fellow Magdalen College, Oxford, Lecturer on Physiology, St. George's Hospital, London.

Nominated by the Royal College of Surgeons—
ALFRED ANTUNES KANTHACK, Esq., B.A., B.Sc., M.B., B.S. (London), F.R.C.S. (England), and L.R.C.P. (London), late Clinical Assistant Royal Ophthalmic Hospital, London, and Midwifery Assistant, St. Bartholomew's Hospital, London.

PROGRAMME OF EXECUTIVE.

Appointed by the Viceroy of India—
(The late) Surgeon-Major BARCLAY, Secretary to the Surgeon-General to the Government of India.
Surgeon-Major S. J. THOMSON, Deputy Sanitary Commissioner of the Second Circle in the North-West Provinces and Oudh.

The following order appeared in the *Indian Government Gazette* of the 21st November, 1890:—

EXTRACT from the Proceedings of the Government of India in the Home Department (Medical), under date, Calcutta, the 21st November, 1890.

RESOLUTION.

The Executive Committee of the National Leprosy Fund having determined to appoint a Commission for the purpose of investigating the disease of leprosy in India, it is notified for general information that Dr. Beaven Rake, Mr. Kanthack, and Dr. Buckmaster (their full titles and qualifications are set out in the resolution) have been appointed Commissioners. These gentlemen have now arrived at Bombay, and have been joined there by Surgeon-Major A. Barclay, M.D., Secretary to the Surgeon-General with the Government of India, and Surgeon-Major S. J. Thompson, Deputy Sanitary Commissioner, North-Western Provinces and Oudh, who have been appointed by the Government of India to co-operate with them. The Governor-General in Council will feel obliged if public bodies and individuals desirous of producing evidence before the Commissioners will address the Local Government of the Province to which they may belong, in order that arrangements may be made for the presentation of such evidence before the Commissioners. Each Local Government has already been requested to depute an officer to assist the Commission in collecting and arranging the evidence that may be procurable in the territories under its control, and instructions should be given to all the civil and medical officers to give to the Commissioners any aid which they may ask for in the course of their inquiries. The Government of India will, on learning from the Commissioners the programme which they intend to follow, notify it for public information.

ORDER.

Ordered, that a copy of this resolution be forwarded to Local Governments and Administrations, in continuation of the communication from this office, No. 11 Medical, 596-605, dated 15th September, 1890, to the Surgeon-General with the Government of India, and to the Members of the Leprosy Commission for information. It will be convenient if the Commissioners will address the Surgeon-General with the Government of India on any point on which they may desire to receive further assistance or information.

Also, that the resolution be published in the supplement to the "Gazette of India."

(True Extract).

(Signed) C. J. LYALL,
Officiating Secretary to the Government of India.

The Leprosy Commissioners, who left England for India on the 23rd October, 1890, completed their inquiries in the autumn of 1891, and prepared their report; but its publication has been delayed, ostensibly on the ground that the Committee of the National Leprosy Fund were waiting the issue of the Indian census returns as regards leprosy. It appears, however, from the facts disclosed in the following circular, issued in June, 1892, that several of the most important conclusions arrived at by the Commission are strongly objected to.

NATIONAL LEPROSY FUND.

Memorandum on the Report of the Leprosy Commissioners.

Your Committee, having been instructed to consider and report to you upon the publication of the Report of the Leprosy Commissioners in India, in 1890-91, beg to submit the following considerations :—

I. They desire to place on record their sense of the ability with which the Commissioners conducted their investigations while in India, and of the comprehensive and valuable nature of their Report.

II. The conclusions at which the Commissioners arrived have been summarised by them at the end of their report as follows—the evidence upon which these conclusions rest being displayed at length in the earlier pages of the Report :—

> (1) "Leprosy is a disease *sui generis;* it is not a form of syphilis or tuberculosis, but has striking ætiological analogies with the latter."
>
> (2) "Leprosy is not diffused by hereditary transmission ; and for this reason, and the established amount of sterility among lepers, the disease has a natural tendency to die out."
>
> (3) "Though, in a scientific classification of diseases, leprosy must be regarded as contagious, and also inoculable, yet the extent to which it is propagated by these means is exceedingly small."
>
> (4) "Leprosy is not directly originated by the use of any particular article of food, nor by any climatic or telluric conditions, nor by insanitary surroundings ; neither does it peculiarly affect any race or caste."
>
> (5) "Leprosy is indirectly influenced by insanitary surroundings, such as poverty, bad food, or deficient drainage or ventilation ; for these, by causing a predisposition, increase the susceptibility of the individual to the disease."
>
> (6) "Leprosy, in the great majority of cases, originates *de novo*, that is, from a sequence or concurrence of causes and conditions, dealt with in the report, and which are related to each other in ways at present imperfectly known."

III. Thirdly, the Commissioners having been instructed to report upon the practical measures to be taken for the control or restriction of the disease in India, have suggested the regulation of lepers and leprosy by means of bye-laws framed by the various municipalities, upon which point they write as follows :—

> (a) "The Commission are of opinion that the sale of articles of food and drink by lepers should be prohibited, and that they should be prevented from practising prostitution, and from following such occupations as those of barber and washerman, which concern the food, drink, and clothing of the people generally, quite apart from the dread of a possible infection."
>
> (b) "The Commission consider that the best policy in dealing with the concentration of lepers in towns and cities is to discourage it, and to this end would suggest that municipal authorities be empowered to pass bye-laws preventing vagrants suffering from leprosy from begging in or frequenting places of public resort, or using public conveyances."
>
> (c) "The large presidency towns and the capitals of provinces in many cases already possess leper asylums, which might be enlarged by municipal funds or private subscriptions. Asylums should be built near towns where they do not already exist, and the authorities should have the power of ordering lepers infringing the regulations either to return to their homes or to enter an asylum."
>
> (d) "Competent medical authority should always be consulted before action is taken under such bye-laws."

IV. Upon the afore-mentioned conclusions of the Commissioners, numbered 1, 2, 3, 4, 5, and 6, your Committee offer the following remarks :—

They accept Nos. 1, 2, 4, and 5.

They desire to express their disagreement with the concluding words of No. 3—

"That the extent to which leprosy is propagated by contagion and inoculation is exceedingly small"—

not being satisfied with the evidence offered by the Commissioners for this opinion.

They cannot concur in the views expressed in No. 6,— namely, that

* "Leprosy, in the majority of cases, originates *de novo*, that is, from a sequence or concurrence of causes and conditions, dealt with in the report, and which are related to each other in ways at present imperfectly known,"

being of opinion that the evidence adduced does not justify such conclusions.

V. The Commissioners, in the section of their Report entitled "Practical Suggestions," pp. 452-7, as also in other parts of the Report, have expressed opinions strongly adverse to compulsory segregation, either complete or partial. For instance, they say on p. 258—

"No legislation is called for on the lines either of segregation or of interdiction of marriages with lepers."

And on p. 453—

"For India, complete compulsory segregation of lepers may be considered to be absolutely impracticable. Neither do the conclusions given before as to the nature of the disease justify any recommendation for absolute segregation."

And on p. 454—

"It is impossible, for the same reasons, to advise compulsory partial isolation. Voluntary isolation is therefore the only measure left for consideration."

---

* See p. 304.

And on p. 456—

"In no case would the Commissioners suggest an Imperial Act, especially directed against lepers as such."

And again on p. 456—

** "In conclusion, the Commissioners believe, from the considerations and arguments adduced in the foregoing report, that neither compulsory nor voluntary segregation would at present effectually stamp out the disease, or even markedly diminish the leper population, under the existing conditions of life in India."

Your Committee, having already expressed their inability to accept the reasoning upon which the Commissioners have based the above conclusions, are equally unable to accept the corollary that segregation in any case of leprosy in India is either impracticable or undesirable. They entertain a precisely opposite opinion, and would be sorry if the Government of India were encouraged by the report of the Commissioners to refrain from taking the necessary steps in the direction of such segregation of lepers as may be found possible. Their opinions upon segregation are in accord with those expressed in the following extract from a memorandum by Dr. Vandyke Carter:—

MODES OF SEGREGATION.

I. "By erecting plain asylums at certain centres, each of which would be a refuge common to several districts, and a place of detention, under due management and supervision."

II. "By founding leper colonies, or village communities mainly of the affected, who, while allowed more liberty of movement, should yet be prevented from mingling with the peasantry around: hence still the need of strict supervision. Many spots would thus serve—such as deserted forts, decayed villages, and places now waste, yet not far from other sources of supply, or not without near resources easily resuscitated."

** See p. 304.

III. "By requiring the strict isolation of leprous subjects retained in their homes at express wish of friends. Suitable separate lodgment would be indispensable; unsuitable shelter is even now sometimes supplied. Joining of such home-isolation with more public measures should not be overlooked, for to it experience in Norway seems to point as a means essential to complete success within a moderate period of time ; and in India it would have to be still more largely resorted to."

IV. "For carrying out the above, in addition to funds, legislative authority is needed to take up the vagrant sick, to remove the sorely diseased who is insufficiently guarded at home, and at times to enforce continued isolation of the infected until medical sanction of liberty be granted."

VI. Reserving their opinions as expressed in the foregoing paragraph and extract, your Committee give a general approval to the minor recommendations of the Commissioners, numbered above as *(a) (b) (c) (d)*, for the regulation of lepers and leprosy in India, which they consider might with advantage be carried out; though they do not concur in the opinion that municipalities will be necessarily or universally the best means of effecting that object.

Nominated by the Executive Committee of the National Leprosy Fund—

GEORGE N. CURZON (Under-Secretary for India), *Chairman.*
EDWARD CLIFFORD.

Nominated by the Royal College of Physicians—
DYCE DUCKWORTH, M.D., LL.D.
G. A. HERON, M.D., F.R.C.P.

Nominated by the Royal College of Surgeons—
JONATHAN HUTCHINSON, LL.D., F.R.S. *(With the exceptions noted below.)*
N. C. MACNAMARA, F.R.C.S.

* Upon this paragraph Mr. Hutchinson appends a dissentient opinion, as follows :—

"I understand the Commissioners to mean by the expression '*de novo*' in reference to the origin of leprosy, that they believe that the disease may begin independently of personal contagion and in connection with climatic and dietetic causes. In that belief I entirely share. I also agree in the main with the rest of the statements in the Commissioners' Report to which exception has been taken in our Committee. I feel convinced that if leprosy be contagious at all, it depends but to an almost infinitesimal extent upon contagion for its spread."

JONATHAN HUTCHINSON, LL.D., F.R.S.

** Upon this paragraph of the Report, Sir Dyce Duckworth and Mr. Hutchinson append independent or dissentient opinions, as follows :—

"I am in agreement generally with the recommendations of the Commission respecting *voluntary isolation*, and the issue of *Municipal Bye-Laws* regulating the habits of lepers. I know of no trustworthy evidence to prove that a leper in any community is a source of greater danger than is a consumptive patient, and I know that a person suffering from syphilis is a real and very positive source of danger anywhere. It would therefore be absurd on the face of it to adopt stringent laws for the leper and to let the syphilitic person go free.

"The intelligent layman now imagines that because bacilli are an essential feature of leprosy, therefore the disease *must* be readily contagious. This is simply quite contrary to fact. The same thing holds good exactly for consumption.

"I think a well-empowered and vigorously-supported Government Medical Executive Officer should be appointed in every large town, and in certain districts, to supervise the leprous populations and report regularly upon them. It should be his business to see that the local regulations are fully carried out, and on his requisition only should any action be taken when necessary.

"Suitable asylums should be provided, and those now existing be sufficiently enlarged to meet the needs that will arise under suitable bye-laws.

"The project of leper-farms is, I think, a good one. More than this is, I believe, not within any practical scheme for amending the condition of lepers, and for diminishing the spread of the malady."

DYCE DUCKWORTH, M.D., LL.D.

"I am strongly in favour of the maintenance (by Government or otherwise) of voluntary homes for lepers, but do not believe that segregation would effect anything in diminishing the prevalence of the disease. Compulsory segregation would I think involve injustice and entail much social misery. I believe that our Commissioners' Report well expresses not alone the opinions of those who have signed it, but, in a general way, those of the educated classes of the present day throughout India."

JONATHAN HUTCHINSON, LL.D., F.R.S.

ADDITIONAL REPORT.

"Your Committee having also been instructed to report upon the disposal of the balance of £800 still remaining to the account of the Executive Committee, recommend that a sum of £250 should be set apart for the prosecution of further investigations and the continued half-yearly publication of the journal for a period of five years, or until such time as this sum is exhausted; and that the remaining £550 should be devoted to the encouragement of local research in countries where there is reasonable evidence for believing that leprosy has recently originated, or where it exists under very exceptional circumstances—with the view of tracing the disease to its alleged origin."

It may be observed that for the Commissioners to have allowed that leprosy is easily inoculable (as is shown in this volume upon the evidence of accepted authorities in all parts of the world) would have been equivalent to the admission of the danger in all leprous countries of the invaccination of leprosy. Such an avowal would have been inconsistent with the course adopted by the Commissioners and by the Committee of the National Leprosy Fund during the whole of this important inquiry, both of whom have practically ignored the evidence bearing upon the subject. When visiting Calcutta, for example, nothing would have been easier than for the Commissioners to have investigated the cases circumstantially reported by Dr. Roger S. Chew, which will be found in another chapter of this book. Having ignored these cases, one of the Commissioners, Dr. Beaven Rake, immediately on his return from India, gave evidence before the Royal Commission on Vaccination to show that leprosy was not communicable by vaccination, or, if so, only to so slight a degree that the danger might be disregarded. How far Dr. Rake's testimony has stood the test of cross-examination will be gathered by reference to the evidence in the Blue Book which had not been issued at the time of this writing.

On the subject of inoculation, the Commissioners conclude :—" The extent to which leprosy is propagated by contagion and inoculation is exceedingly small." It is a pity that the Commission should have bundled these two dissimilar sources of alleged causation together, as they cannot consistently be so treated. Nor have they defined the word "contagion," which many authorities, as I have shown, habitually use to cover inoculation. The dissemination of leprosy by *contagion*, using the word in its proper sense—*i.e.*, by simple contact—has been disproved by eminent authorities, including superintendents of leper asylums of wide experience in all countries, as will be seen by reference to the chapter in this book entitled " Is Leprosy Contagious ? " On the other hand it is conclusively established by a similar weight of evidence that leprosy is inoculable, and like other transmissible diseases can be propagated through a cut, sore, wound, or abraded surface, or be inoculated by flies and mosquitos, or spread by vaccination. Another point in which the Committee dissent from the views of the Commissioners is that of compulsory segregation or compulsory isolation, which they consider uncalled for by the evidence collected during their tours of investigation, or by the nature of the disease, and are of opinion that it would be impossible to stamp out the disease by either voluntary or compulsory segregation. The Committee of the National Leprosy Fund do not accept either the reasoning or conclusions upon which these recommendations of the Commissioners are founded. On the contrary, they are of the opinion that the segregation of lepers should be encouraged as far as possible in the interest both of the lepers and of the public. The Committee conclude by declaring that they can give only "a general approval to the minor recommendations of the Commissioners" on this point. Both the Committee and the Commissioners accept the important fact that leprosy is not diffused by hereditary transmission.

Setting aside the alleged leading causes of leprosy, such as heredity, contagion, and inoculation, as contributing little or nothing to the spread of this disease, the

Commissioners resort to a theory which, up to the present time, has received but little countenance from the medical profession, that "leprosy in the great majority of cases originates *de novo*, that is from a sequence or concurrence of causes and conditions, dealt with in the Report." The present writer considers it highly probable that leprosy as well as other diseases may in numerous instances be accounted for in this way, but it must be observed that the factors, which are various forms of insanitation, have been in operation time out of mind, and do not in any way account for the remarkable recrudescence of the disease shown in this volume. On the contrary the danger from malaria, overcrowding, impure water, unwholesome food, filthy deposits, has, under the instructive teachings of the late Dr. Southwood Smith, Sir Edwin Chadwick, and their followers, been gradually and sensibly diminishing, and it is triumphantly claimed that various diseases due to these causes have been decreasing also. It is obvious, therefore, that some other factor or factors, peculiar to this century and previously unknown, are at work. This factor, as high authorities now reluctantly avow, has been omitted by the Commission from the list of causations in their summary of conclusions—the latest and most daring official effort, to use a classic phrase, "to preserve vaccination from reproach."

Amidst this divergence of opinion between the conclusions of the Leprosy Commissioners and those of the Committee, the public will want to know what useful object has been realised by the large expenditure of time and money in promoting this lengthened inquiry. Will the authorities accept the conclusions of the Leprosy

Commission, when they affirm that "the extent to which leprosy is propagated by contagion and inoculation is exceedingly small;" or the opposite views entertained by the eminent members of the Committee of the National Leprosy Fund, who declare their disagreement with these amongst others of their conclusions?

The Leprosy Commissioners (all ardent supporters of the Jennerian practice) have searched far and wide for a rational theory that will account for the recent spread of leprosy in certain countries, but have utterly failed to discover one, and are almost driven to the conclusion that touches closely upon the facts collected in this volume. Under the head of "Transmission through an Intermediary Host," one of their number, Dr. Beaven Rake, says :—" There is at present no direct evidence to support this hypothesis, but it seems that some such theory might explain the alleged rapid increase of leprosy in Hawaii and New Caledonia."—*Journal of the Leprosy Investigation Committee, No. 1, Aug., 1890, pp. 50, 51.*

To the present writer, vaccination seems to be the only intermediary host that passes direct into the blood (except in rare instances of accidental inoculation), and which covers all the facts of the case.

It appears that of the £7000 collected £800 remains, part of which it is suggested should be devoted to further investigation. It is to be hoped, if this is adopted, that less prejudiced and more competent inquirers will be appointed, and that in the future numbers of the *Journal* the accumulation of evidence, so shamefully ignored, showing how leprosy has been spread at the point of the vaccinator's lancet will find a place.

CHAPTER XVII.

LEPROSY INCURABLE—HYGIENE THE ONLY PALLIATIVE.

LEPROSY has been regarded in every age and in every country as an incurable disease.

In the *Encyclopædia Britannica*, leprosy is described as an incurable constitutional disease, marked externally by discoloured patches and nodules on the skin, and deeply implicating the structure and function of the peripheral nervous system.

How improbable was all hope of cure of leprosy may be inferred from ancient customs in various countries, notably in France.

Dr. Macnamara in his work on " Leprosy," p. 36, cites authorities to show that the leper was expelled from society, and looked upon as dead.

He observes:—"The leper was not looked upon in the eye of the law alone as defunct, for the Church also took the same view, and performed the solemn ceremonials of the burial of the dead over him on the day on which he was separated from his fellow-creatures and consigned to a lazar-house. He was from that moment regarded as a man dead amongst the living, and legally buried, though still breathing and alive. The ritual of the French Church retained till a late period the various forms and ceremonies to which the leper was subjected on the day of his living funeral. Ogee and Ploucquet have both described them.

"A priest robed with surplice and stole went with the cross to the house of the doomed leper. The minister of the church began the necessary ceremonies by exhorting him to suffer, with a patient and penitent spirit, the incurable plague with which God had stricken him. He then sprinkled the unfortunate leper with holy water, and afterwards conducted him to the church, the usual burial verses being sung during their march thither. In the church the ordinary habiliments of the leper were removed; he was clothed in a funeral pall; and while placed before the altar between two trestles, the *Libera* was sung, and the mass for the dead celebrated over him. After this service he was again sprinkled with holy water, and led from the church to the house or hospital destined for his future abode. A pair of clappers, a barrel, a stick, cowl, and dress, etc., etc., were given to him. Before leaving the leper, the priest solemnly interdicted him from appearing in public without his leper's garb; from entering inns, churches, mills, and bake-houses; from touching children, or giving them aught he had touched; from washing his hands, or anything pertaining to him, in the common fountains and streams; from touching in the markets the goods he wished to buy with anything except his stick; from eating or drinking with others than lepers; and he especially forbade him from walking in narrow paths, or from answering those who spoke to him in the roads and streets unless in a whisper, that they might not be annoyed with his pestilent breath, and with the infectious odour which exhaled from his body; and last of all, before taking his departure and leaving the leper for ever to the seclusion of the lazar-house, the official of the church terminated the ceremony of his separation from his living fellow-creatures, by throwing upon the body of the poor outcast a shovelful of earth, in imitation of the closure of the grave."

Referring to more recent events, I find that in the report of the Special Sanitary Committee on the state of the Leper Settlement at Kalawao, Hawaii, 1878, addressed to the Hon. G. Rhodes, President of the Legislative Assembly, so little expectation of cure was there, "that lepers have to pay for their own coffins, and have formed a coffin association in order to provide a common fund for their proper interment, and these sad creatures get up, as shown by the register of the Hospital, 'Coffin Feasts,' on which occasion money is contributed to provide for a decent termination of their woes."

Alluding to the disappearance of leprosy in England, Gilbert White observes in a letter to Mr. Barrington:—
" This happy change perhaps may have originated and been continued from the much smaller quantity of salt meat and fish now eaten in these kingdoms; from the use of linen next the skin; from the plenty of better bread; and from the profusion of fruits, roots, legumes, and greens, so common in every family." It may also be added that, at the time when leprosy disappeared from this country, the practice of inoculation and vaccination was unknown, otherwise there is little doubt that leprosy would have been perpetuated in England by the empoisoned lancet, as it is now in the West Indies, British Guiana, India, New Caledonia, the United States of Colombia, Venezuela, and Hawaii.

Commenting upon the Royal College of Physicians'

Report, Drs. Tilbury Fox and Farquhar observe that: "The cause of leprosy is as obscure as ever, and upon this particular matter the leprosy report gives us very little satisfactory explanation, beyond illustrations of the general statement that leprosy disappears *pari passu* with an improvement in the hygienic condition and diet of a people, and the cultivation of land in districts where it has abounded."—*India Office Report, London, 1872, p. 28.*

Sir Erasmus Wilson, in his article on "Leprosy" in Quain's "Dictionary of Medicine," refers to the various drugs which are recommended and used by one physician or another—quinine, strychnine, phosphates, nitric acid, acetic and carbolic acid, iodine, arsenic, perchloride of mercury, asclepias gigantea, hydrocotyle asiatica, veronica quinquefolia, plumbago rosea, acid nitrate of mercury, potassa fusa, acrid irritating oil of the shell of the cashew nut, chloride of zinc, etc. In no case is it mentioned that a cure is effected, or even to be expected, but rather the other way. "Hope," he says, "will gleam in the mind of the physician and patient; but cure, alas! is as distant as ever."

Dr. P. Abraham, the Secretary of the Leprosy Investigation Committee, who is familiar with all medical literature on the subject, gave his opinion at a discussion "On the Cure of Leprosy" before the Royal Medical and Chirurgical Society, May 27, 1890, that "the disease was probably not curable, though it might abort and die out."—(*British Medical Journal*, May 31st, 1890.) And on another occasion he observed that therapeutic agents had proved failures and that there was no specific for the disease.

In a report of the Paris Congress of Dermatology, 1889, printed in the *Journal d Hygiène*, October 23rd, 1890, the subject of leprosy was discussed.

M. Cornil observed: "We must not confound a passing amelioration, a diminution or attenuation of the malady, more or less prolonged, with cure properly so called."

Consul-General Abbott, in a letter dated June 8th, 1891, Rio de Janeiro, says: "I do not think any radical cure for leprosy has yet been discovered."—*Journal of the Leprosy Investigation Committee, No. 4, December 1891, p. 18.*

HAWAII.

The health authorities in Hawaii, as well as the most experienced medical residents elsewhere, consider leprosy as practically incurable, though they acknowledge that life may be prolonged by good food, pure water, healthy habitations, and other favourable sanitary conditions. On one occasion, referring to the reports of certain medical practitioners as to their alleged and officially reported cures, I was told by one of the responsible officials that the cures were mythical, the particular cases cited having since exhibited further development of the disease.

It is alleged that the external manifestations of leprosy, like those of some other maladies, disappear for a time, either with or without treatment, and cures are noised abroad; but the disease invariably reappears, and shows itself when least expected.

In the year 1866, the disease had begun to assume alarming proportions in the Sandwich Islands, and the people strongly suspected that the increase was partly, if not largely, due to vaccination. I am unable, however, to find any medical admission of the fact, until some years later. A law was passed in 1868, declaring the disease to be contagious, and requiring all lepers to be removed to the Island of Molokai, where the Government had set aside a piece of land on the northern side for a leper settlement. The preparations for the reception of these unfortunate people were, at first, of the rudest description. The patients were lodged in the cottages of the few Kamaainas, or freeholders on the estate, and were without any appropriate nursing or attendance; and, according to the evidence of Mr. Ambrose Hutchinson, who in 1884 was Under-Superintendent, as the disease was considered incurable, it was the custom to send along with each patient, by the same conveyance, the coffin he was soon to occupy.

The report of the Honolulu Board of Health to the Minister of the Interior for 1876 states (page 77) that the Legislature of 1874 made an appropriation of six thousand dollars towards the expenses (for curing leprosy) of Drs. Powell and Akana, who were offered every facility to try their skill. Six patients were experimented upon with various drugs. Dr. Akana claimed to have cured one, who was subsequently examined by a number of physicians and sent to Kalawao as a confirmed leper. The Board reported that all attempts to cure any patient afflicted with leprosy had failed.

In an address delivered by Mr. Dole on the subject of "Leprosy in the Hawaiian Islands," which the *Boston Medical and Surgical Journal* of May 15, 1884, commends to the attention of the National Board of Health, United

States, is the following description of this malady:—
"Leprosy is the worst known disease of the present and historic times. It has successfully defied medical skill. Physicians have not been able to say whence it comes, or to explain its laws. It has always and everywhere been found to be incurable. It attacks all races and all classes; no rank in life is safe—adults and children are alike exposed to its ravages. Medical skill can make no limit; no assurance can be given from the lessons of experience against this most terrible, most loathsome, and most hopeless of all human diseases."

Dr. A. W. Saxe, in a card published in the *Hawaiian Gazette* of May 23, 1883, says that for leprosy there is no known cure.

In a paper on "Leprosy in Hawaii," published in the *Occidental Medical Times*, April, 1889, Dr. F. B. Sutliff, the writer, sums up the results of his experience in the Island of Maui, Hawaii. "Treatment," he observes, "of any kind has, so far, proved useless. Improvement which has been noted in many cases is only temporary;" and again, p. 208, "The disease is hopelessly incurable, and certainly fatal."

In the biennial period ending 1881, the report of the Board of Health states that an appropriation of $20,000 was made by the Legislature of Honolulu for the cure of lepers, or $200 each; but no cures were effected, though numberless experiments on these unfortunates were tried, causing much suffering, and in one case the suicide of the victim experimented upon.

In his report to the Board of Health, dated November 14, 1885, and addressed to His Excellency W. M. Gibson, Dr. Edward Arning observes:—" There is scarcely a drug in the pharmacopœia, at least scarcely a class of drugs, that has not been most systematically tried in the treatment of leprosy. Over and over again men of sanguine temperament have found what they called a specific cure, but in every instance calm and unbiased judgment has afterwards pronounced a verdict of uselessness."

This distinguished authority also observes:—" I am fully satisfied that Dr. Unna has as yet not succeeded in perfecting a cure of leprosy in a single case. He himself has told me that one of the published cases of complete cure has since suffered a relapse."—*Biennial Report, Board of Health, Honolulu, 1892, p. 5.*

Leprosy is not confined, as many suppose, to native Hawaiians. Dr. E. Arning says:—" Among the white population, numbering 17,935, I knew of thirty-five leprosy cases."

Dr. N. B. Emerson, a lepra specialist of Hawaii, says: —" In spite of a number of claims to the contrary, we believe it safe to say that no one has been able to prove, to the satisfaction of the medical profession, that a single case of this disease has been definitely cured."

In a volume entitled " Leprosy in Hawaii," published by the Hawaiian Government at Honolulu, 1886, p. 73, it is said:—" Experiments are carried on constantly on patients at Kalihi, and on very incipient, or rather doubtful cases, at their own homes. Though some patients have certainly improved a great deal under careful treatment, we cannot, for the present, state one case of cure."

Dr. Prince A. Morrow, of New York, in " Personal Observations of Leprosy in Mexico and the Sandwich

Islands," says, p. 5 :—"We know that leprosy has a prolonged, but somewhat indefinite, period of incubation, a slow and irregular course of development, a characteristic and well-defined symptomatology rendering its diagnosis easy, and that its prognostic significance is most grave. It progresses almost invariably to a fatal termination."

Dr. Hoffman, of Honolulu, alluding to the Leper Hospital at Kalihi, of which he had charge for some years, reports that generally fifty cases were under treatment from time to time, these being equally divided between the tubercular and the anæsthetic. He says :—"I found no permanent benefit from treatment; better food and cleanliness, medicine suited to improve the general health of the leper, ameliorated the disease temporarily."—*Report on Leprosy*, *Honolulu*, *1886*, *p. 75.*

Dr. Arthur Mouritz, Resident Physician and Medical Superintendent of the Leper Settlement, Molokai, in his Report, dated Molokai, February, 1886, says of the Kalawao Hospital, where the worst cases are supposed to be accommodated:—"Of course, there are no cases of cure, and those who enter its portals remain until death releases them;" and under the head of "Treatment," says: "This is the briefest question of any to deal with, but the most disheartening to a physician; for, so far, no remedy has been found beneficial." Dr. E. Cook Webb, physician to a branch hospital at Kakaoko, in his report, dated March 1st, 1886, says:—"As regards the treatment of leprosy, I have but little to say. Notwithstanding any treatment which I have used, or seen used, I cannot see any change in any single case. I am fully convinced, after considerable study and experience, that personal cleanliness, good nourishing food, and regular habits have done more towards the relief of these unfortunates than all the medicines that have ever been prescribed for them in the past. In all the cases of leprosy I have seen, the disease has steadily progressed to a fatal termination, notwithstanding all treatment. I am aware that I am taking strong ground against the many so-called 'cures' that have been devised, but in so doing I am not basing my opinion on my own study and experience alone, but on the medical opinions of those who for years have been in daily contact with the disease, and have made it a special study, and they have come to the conclusion that it is a disease *sui generis*, and incurable."

In the report of the President of the Board of Health, addressed to the Legislative Assembly, Honolulu, 1886, p. 49, in reply to the question, "Is Leprosy Curable?" Mr. F. W. Hutchinson says :—" To this question we are constrained to answer 'No!' At least not under any known treatment."

Dr. Sidney Bourne Swift, late Resident Physician of the Leper Settlement, Kalawao, Molokai, Hawaii, in his report to the President of the Board of Health, refers to the failure of medical treatment and the numerous alleged cures reported by experimental bacteriologists. "What has become of the cases cured by Fitch, by Arning, by Goto, and by Lutz? Come to the Settlement and you will see them in their graves."—*Biennial Report of the President of the Board of Health to the Legislature of the Hawaiian Kingdom, Honolulu, 1892, p. 87.*

THE WEST INDIES AND BRITISH GUIANA.

Dr. Gavin Milroy, in the "Report on Leprosy and Yaws in the West Indies," published in 1873, p. 43, observes:—"We have already seen that in the Barbados Asylum several of the patients had been benefited when no medicinal remedies whatever had been used, the good results being attributed solely to the more regular mode of life, better supply of food, and better housing since their admission, and to their withdrawal from all occasions of intemperance on the one hand, and of destitution on the other. Like effects have been observed in the asylums of other Colonies."

On pages 99 and 100, Dr. Milroy says:—"There is a unanimous accord of opinion that the greatest benefit is derived from the adoption of hygienic measures, and that by improving the general conditions, physical and moral, of the leprous poor, very much may be done to retard or arrest the malady in its early stages, and also to mitigate its severity when more fully developed. Medicinal treatment is universally admitted to be of no avail, unless combined with the regular use of a nutritive, unstimulating diet, suitable clothing, protection against the vicissitudes of weather, personal cleanliness, and exercise in the open air. There is certainly no medicinal substance, vegetable or mineral, which exerts anything like a direct or specific effect on the malady. . . . .

"The evidence is all but unanimous that leprosy very rarely, if ever, manifests any tendency to a spontaneous cure. When fully developed, a complete recovery is not to be looked for. It is quite apparent, however, that the progress of the disease may often experience a marked retardation of arrest when the patient is maintained in a favourable hygienic condition."

Dr. J. F. Donovan, Superintendent, Leper Home, Jamaica, in his annual report for 1891, says:—" For curative purposes, I think most authorities will acknowledge that we must look for some other method besides the administration of medicinal remedies, as drugs have been tried extensively and persistently, and have so far signally failed of effecting a cure of this formidable pest."—*Supplement to Jamaica Gazette, 18th June, 1891.*

In the same report, Dr. Donovan makes the following statement as to the condition of patient No. 5, afflicted with tubercular leprosy:—" After medical treatment, the scaly eruption about the flexor aspect of joints has disappeared, the plagues of tubercle on face, forehead, etc., and the enlarged inguinal glands are in *statu quo* after a month's steady use of the inunction and mixture. Symptoms of pulmonary phthisis set in when the gurjun oil was discontinued, as it caused nausea, and subsequently diarrhœa ensued, which the patient attributed to the medicine. A variety of drugs were used to combat the hyper-pyrexia, but without effect."—*Supplement to the Jamaica Gazette, June 18th, 1891, page 81.*

Under the head of "Cure," Dr. J. F. Donovan observes in the same report:—" In some cases the tubercles decrease for a time; in a few years, it may however be mentioned, that the tubercles have been known to subside spontaneously in some patients for a time, who have been taking no medicines; so too the tubercles may enlarge and increase in numbers during the administration of the drug, as was evidenced in three of our most prominent cases at the 'Home.'"

"There are few diseases which are less amenable to treatment than leprosy; the entire armoury of known drugs has been tried," says this medical superintendent, "and found ineffective."

I may here observe that Dr. Koch's much vaunted tuberculin has been tried in almost every leprous country on the globe, and found worse than useless. Much suffering has been caused by its use. The *Madras Times*, October 28, 1891, reports the visit of His Excellency Lord Wenlock to the leper hospital in this city, when his lordship had pointed out to him several patients upon whom Dr. Koch's lymph treatment had been tried. One of the patients complained that the experiment had subjected him to excruciating pain.

Dr. W. V. M. Koch, Acting Medical Superintendent, Leper Asylum, Trinidad, says:—"During the past year there has been scarcely any advance made in the treatment of this disease (leprosy) and it continues to baffle the most skilful physician. Various new remedies have been tried, but without success."

"No drug has yet been used which exerts a specific action on leprosy."—*Appendix to Report on Leprosy and the Trinidad Leper Asylum, 1891, pp. 72, 73.*

In the appendix to Dr. Beaven Rake's report on "Leprosy and the Trinidad Leper Asylum," dated Maraval, February 4th, 1890, it is admitted that the inoculation of animals has proved a failure, and human beings are required as clinical material for experimental purposes. Dr. Rake, p. 36, says:—"If the Home Government could see its way to sanction the inoculation with leprosy of two or three condemned criminals in Trinidad, and the commutation of their capital sentences to imprisonment for life, important additions could be made to our present knowledge of the pathology and proper treatment (by segregation or otherwise) of the disease. I can safely predict that many criminals would gladly accede to such an alternative, on having the case clearly stated to them."

Dr. Beaven Rake, himself a great experimenter, says:— "Campana inoculated lepers with erysipelas, with the result that nearly all the patients in the ward got erysipelas, and the ward had to be closed. No effect was produced on the progress of the leprosy."—*Report on Leprosy in Trinidad for 1890, p. 37.*

Dr. John D. Hillis, of Demerara, observes in his work, "Leprosy in British Guiana," p. 209:—"The treatment of leprosy has hitherto been attended with such very poor results that the disease is now regarded as incurable. Drug after drug—so-called specifics—have been tried, only to be laid aside as useless, the disease after a time returning with greater violence than ever." The writer then proceeds to describe some of the experiments to which the unfortunate lepers have been subjected. Out of seventy cases treated by Dr. Danielssen at the Lungeguard's Hospital, Bergen, only one was reported cured. "It appeared to me, however," says Dr. Danielssen (page 210), "if I could infect the leprous patients with constitutional syphilis, it might follow that the syphilitic poison might prove superior to that of leprosy, and that thus the system might be brought to that of a person labouring under constitutional syphilis, and might so become subject to the ordinary process of syphilisation." "This ingenious theory, however," remarks Dr. Hillis, "failed in practice, the leprosy remaining unchanged, whilst the syphilitic process went on."

Under the head of "Palliative Treatment," p. 215, Dr. Hillis says:—"Improving the sanitary condition of the leper, it is well known, has great influence in mitigating the disease; and the satisfactory results which have been realised in this direction at the General Leper Asylum may be seen in the following figures, taken from my official report for 1878, when the percentage of deaths to strength for the past four years was stated to be :—

1875.	1876.	1877.	1878.
17·36 per cent.	16·33 per cent.	11·49 per cent.	9·19 per cent.

In the appendix to the report of the Medical Officer of the Leper Asylum, Mahaica, British Guiana, for 1880, printed by order of the Court of Policy, Dr. Oscar D. Honiball, Acting Medical Officer to the Leper Asylum, after alluding to the prominent physiological action of gurjun oil, which are purging and vomiting, considers it would be a fatal mistake to administer it in advanced cases. Dr. Honiball records with considerable regret and disappointment his failure to discern its alleged beneficial results. He has also made careful inquiries of the inmates, and the answer as to the beneficial results have been either of a negative nature or strongly adverse. Some say, 'I am not worse than when I came in.'" On the other hand, very many bitterly complain of the deleterious qualities, attributing to its administration a rapid and violent increase of the disease. It is so nauseous, and the results following so very often serious, and at all times disagreeable and inconvenient, that the assumption of its curative properties should be based upon a surer basis than hypothesis."

Dr. Castor, Medical Director, Leper Hospital, Demerara (who acknowledges that vaccination is a certain mode of spreading leprosy) says that no therapeutic agent is of any avail as a cure.

Referring to Dr. Castor's report to the Surgeon-General, British Guiana, for 1888, the *Lancet*, March 8, 1890, p. 566, observes:—"Although every remedy reported successful elsewhere has been tried, no beneficial result has been obtained. Dr. Castor holds the opinion that the most that can be done in the way either of cure or prevention is 'by proper diet, dwellings, and sanitary surroundings to ameliorate the symptoms, and often thereby control them.'"

AMERICA AND CANADA.

In a chemical lecture on "Anæsthetic Leprosy," by Professor James Nevins Hyde, in the *American Practitioner*, February, 1879, the author says:—"Needless to say that mercury, iodine, quinia, arsenic, and a long list of other remedies, have utterly failed to eradicate the disease. A careful study of the results said to have been obtained by the use of gurjun oil, employed in the Beauperthuy method by Dougal, and of the cashew-nut by De Valence, will lead to the conviction that the benefit was largely due to the improved hygienic condition of the patients submitted to experiment. Where we are ignorant it is best to admit the fact; for we thus show that we have at least learned the alphabet of wisdom."

Dr. J. E. Graham, of Toronto, in reply to a communication from the Hawaiian Government regarding leprosy in New Brunswick says, that he has experienced " no good results from medical treatment. Much may be done by attending to the general health of the patients." *Leprosy in Foreign Countries, Honolulu, 1886, p. 158.*

Dr. H. S. Orme, President of the State Board of Health, California, in his valuable papers on "Leprosy, its extent and control," says:—"The general testimony is to the effect that any mode of treatment is disappointing. Arrest of progress is only temporary, being usually followed by suspension of treatment. Indeed it is not certain that long perseverance would be attended by permanent relief. At the Tracadie Hospital, patients have been discharged apparently cured, but they generally returned to die. The results are even less encouraging than the treatment of pulmonary consumption. The Health Authorities of the Hawaiian Islands consider leprosy practically incurable; though they acknowledge that life may be prolonged by certain medical treatment, by good food, and by favourable sanitary conditions."

Dr. James H. Dunn, Professor of Dermatology, Mineapolis, in a chemical lecture reported in the *North-Western Lancet*, March 1, 1888, says :—" The treatment of leprosy is accordingly largely palliative. Of course, cures—popular, medical, and secret—have not been wanting; but their unreliability has been repeatedly demonstrated. Of prime importance, if possible, is the removal of the patient to a country or part of the country in which the disease is not endemic, preferably a healthy mountainous district with good air, nourishing food, and every hygienic appointment, and the use of baths and electricity in proper cases."

In reply to a communication addressed by me to the Superintendent of the Lazaretto, Tracadie, New Brunswick, Mr. J. A. Babineau, November 12, 1889, writes :— "Leprosy is considered incurable here, as elsewhere, for all attempts to cure the disease have failed."

Dr. J. C. Taché, Visiting Physician to the Tracadie Lazaretto, in p. 150, "Leprosy in Foreign Countries," says :—"The various and multiplied attempts made at different times in New Brunswick by medical men, or under medical guidance, to cure the disease have all failed."

Dr. A. C. Smith, writing from New Brunswick to the Hawaiian Government, says :—"I have never observed more than a temporary amelioration from any medical treatment, and only such as might be attributed to the effect of mind over body. My predecessor used coloured water, accompanied by strong assurances of benefit therefrom, and in every instance found a temporary improvement equal in degree to any apparent benefit to be found from the use of medicinal agents." "Leprosy in Foreign Countries," p. 156.

United States Consul James W. Siler, of Cape Town, South Africa, in the report to his Government, after referring to various so-called remedies, observes :—"After all, these and other remedies only tend to prolong the disease ; for, once affected with the leprous taint, the victim is doomed to slowly but surely rot away, until mercifully released by death."—*United States Consular Report for 1887, p. 565.*

Owing to the remarkable spread of leprosy in Venezuela, the United States Consul, Mr. E. H. Plumacher, has paid much attention to the treatment of that disease; and he reports to his Government (No. 119, for 1890, pp. 695-6) as follows:—"Various methods of treatment have been tested at the Maracaibo Lazaretto, especially the administration of the oil of chaulmoogra, which apparently gives encouraging results at first, but produces no lasting benefit. Its use is also attended with grave physical inconvenience, such as a morbid state of the liver; intestinal irritations, accompanied with a slimy and waxy diarrhœa; wandering pains in various parts of the body; eruptions upon the skin, principally attacking the hands, and various other unpleasant attendant symptoms, making the use of chaulmoogra at times intolerable." . . . This treatment has always been entirely voluntary, but no one has been able to persevere in its use. The iodides and mercurial preparations have also been tested, as well as the tincture of cantharides; but all of these remedies produced more or less identical effects— that of a temporary amelioration of the condition of the patient, but without well-founded hopes of anything approaching a genuine cure. . . . As already stated, temporary alleviation has frequently been obtained by various methods, as, in a disease like leprosy, any remedy which tends to improve the state of the blood and the general health will, no doubt, have its temporary ameliorating effect upon the malady itself. So many years of careful study, and of patient and conscientious application of all methods of treatment, have, in my opinion, satisfactorily demonstrated the incurability of the disease; and the most that can be done is to alleviate, as far as possible, the physical suffering and mental distress."

Miguel Valladores, Physician of the Lazaretto, Guatemala, says :—" I have observed that mercurial treatment aggravates the disease of leprosy in a patient."—*Leprosy in Foreign Countries: Honolulu, 1886, p. 174.*

INDIA AND OTHER COUNTRIES.

In a report to His Hawaiian Majesty, dated August 12, 1885, by Surgeon-General W. J. Moore, of Bombay, the author says :—" If leprosy is not what I hold it to be, we have still sufficient evidence that the great prophylaxis is sanitation. In sanitation I include the prevention as much as possible of whatever entails a state of human system below par, such as the cheapening of salt (an article of the greatest importance in the human economy), plentiful food, good clothing, suitable and, above all, dry lodging, drainage, conservancy—in short, everything tending to improve the condition of the population of a country. Leper asylums are good and charitable, but will not cure, eradicate, or prevent leprosy. There is no known cure for leprosy when once contracted. Lepers taken into an asylum and well cared for often apparently recover ; but the apparent recovery is this: The cachectic debilitated leper becomes temporarily a robust leper, but he remains a leper still, and the disease eventually breaks out again. Apart from charitable motives, therefore, I would not recommend the Government spending large sums on leper asylums —such, for instance, as would be entailed by a ' State Leper Asylum,' as mentioned in Government Resolution No. 2009, dated 11th June, 1883. A more certain, albeit slow, progress will result from sanitation in the broadest sense of the term, which comprises the moral and natural amelioration of the condition of the people." And in a letter to the *British Medical Journal*, December 14, 1889, p. 1371, Sir William J. Moore observes :—" I feel most confidence in the diminution of leprosy in India, and in the prevention of leprosy in this country, from the influence and progress of sanitation in the most extended sense of the term, in which I include the cleansing generally of villages and towns, drainage, ventilation, good water supply, the cheapening of salt, the prevention of scarcity, opposition to imprudent marriages, and measures for the prevention of specific disease."

Mr. W. Walker, Inspector-General of Civil Hospitals, North-West Province of Oudh, in his report of June 26, 1885, to the Government, observes :—" I may say that medical treatment, in the sense of attempting a cure of the disease, has been abandoned, not only in these provinces, but all over India. Extensive experiments were made in 1875, 1876, and 1887, with regard to the efficacy of certain systems of treatment, and were found to be equally unsatisfactory. If the Government will refer to proceedings in the Medical Department, Nos. 20 and 23, dated March 10, 1887, there will be found the result of a fair trial given to gurjun oil, once a vaunted cure for leprosy. The results of this experiment may be taken as a fair example of the conclusions which have been forced on all trustworthy observers—namely, that good nourishing diet, cleanliness, and friction of the skin with any oil, are the only satisfactory means of retarding the progress of the disease. No other specific treatment is now attempted in any of our asylums. The patients are regarded as incurable, and are only subjected to medical treatment when attacked by complications which may be hopefully dealt with."

J. Fairweather, Brigade-Surgeon, Inspector-General of Civil Hospitals, Punjaub, says:—"All attempts at specific treatment have been abandoned for some years as useless."

Dr. W. R. Kynsey, Chief of the Medical Department, Colombo, Ceylon, in reply to a communication from His Hawaiian Majesty's authorities as to leprosy in India, writes:—"No treatment has yet been found of any permanent benefit. The best results have been obtained from hygienic and dietetic treatment alone." "Lepers," he says, "are chiefly found among the poorer natives, whose dwellings are small thatched huts, crowded, ill-ventilated, filthy, and strewn with mouldy and rotten vegetables and excremental deposits."—*Leprosy in Foreign Countries, Honolulu, 1886, p. 164.*

Mr. H. A. Acworth, Municipal Commissioner, Bombay, says:—"Who ever heard of a case of cure?"

Dr. R. J. Wright, Civil Surgeon, Jessore, India, says: —"Six hundred and nineteen lepers are reported in the Jessore district. The sex is not distinguished; but it appears that only fifty-four inherited leprosy, while six hundred and sixty-five have no idea of its cause. They believe the disease incurable, so it is difficult to persuade them to submit to treatment, and twenty who were treated with gurjun oil derived no benefit."—*Leprosy in Foreign Countries, Honolulu, 1886, p. 34.*

Dr. H. V. Carter, of Bombay, in a communication to the Hawaiian Government, relative to leprosy in India and Norway, dated 1884, says:—"*In limine*, I should state that the cure of leprosy by purely medical treatment has not practically contributed anything towards the obliterating of the disease. To rely, therefore, for a general amendment, upon any of the varied remedial measures often confidently put forward, would be to indulge in fallacies, hurtful as well as deceptive, and to encourage a kind of anticipation hitherto shown by experience to be futile."

Again, the same writer, in part II. of the same paper, observes :—" Nor has purely medical treatment ever proved curative ; and, so far from leprosy in Norway showing a natural tendency to subside, there is ample evidence of a present activity, equal to that displayed by the disease twenty-five years ago."—*Leprosy in Foreign Countries, p. 96, Honolulu, 1886.*

In a summary of his work for 1890, in the Leper Asylum at Dehra Dun, India, Dr. M'Laren, the Medical Superintendent, gives the result of the treatment with resorcin icthyol and gurjun oil, which were used a short time ago, but are now, apparently, abandoned. He says :—"There is not the least appearance of permanent benefit, or of any amelioration of the actual disease."—*Calcutta Englishman, March 7, 1891.*

In his "Remarks on Leprosy," the same authority observes:—"It may not be out of place here to mention that, since 1875, I have given the most careful attention to the treatment of leprosy ; tried most conscientiously all the various drugs that have from time to time been recommended, and used unsparingly, and for prolonged periods, all outward applications that have been brought to notice, and must frankly admit that I have not witnessed the least *permanent* benefit from any one of these. . . . Once a leper, always a leper, is the sad outcome of my many years' close observation, let the treatment be what it may."—*My Leper Friends, by Mrs. Hayes, London, 1891. pp. 125-6.*

Under the head of "Treatment," the chief medical officer of Kashmir, Dr. A. Mitra, after alluding to the use of arsenic, chaulmoogra oil, and gurjun oil, observes that they have little or no power in arresting the progress of the disease. And as regards nerve-stretching, which, in the early stages of the anæsthetic variety, produces very satisfactory results in a large majority of cases, he remarks "the result is not lasting."—*The American Journal of Medical Sciences, July, 1891.*

The Bombay Gazette, 17th July, 1891, reports the opening of the New Leper Asylum at Schore, Bhopal, towards which Her Highness the Begum of Bhopal has contributed munificently, and has promised an annual grant of 4500 Rs. for the expenses of maintenance. The building will accommodate about 160 lepers. In his address, Surgeon-Major Dane frankly said, "We do not expect to cure these unfortunate people, as, notwithstanding the praises which are repeatedly being bestowed on some vaunted 'certain cures' there is no doubt that a cure for leprosy has still to be discovered." This benevolent lady, the Begum of Bhopal, Nawab Shahjeham, has been persuaded to extend vaccinations in her province, upon which she spends 5000 rupees yearly, employing 35 vaccinators, who performed 38,000 vaccinations last year, thus unwittingly spreading the fell disease at the point of the lancet, and helping to fill the wards of the hospital which her benevolence has established.

Babu Prosurmo Coomar Sein, Gurbetta, India, in reply to questions from the Hawaiian Government, reports as follows : — "After taking charge of this dispensary, I have treated twenty lepers. To some of them this disease was hereditary, to some it was owing to the contagion, and to others it was the effect of using mercurial medicines."

Surgeon W. D. Stewart, Civil Surgeon of Cuttack, says :—" There is a belief that excessive use of mercury tends to develop the disease, by causing a deteriorated state of the blood and tissues."

Babu Jaggo Mohoun Roy, Orissa, reports :—"In the generality of the cases, especially among the Hindus, venereal diseases, and perhaps administrations of mercury for their cure, have, I believe, been the cause of leprosy."
—*Leprosy in Foreign Countries, Honolulu, 1886, pp. 33, 37, 39.*

Surgeon-Major Geoffry C. Hall, Allahabad, in a communication to the *Indian Medical Record*, says :— "Is there any cure for leprosy? I reply, No. Can leprosy be mitigated? I say decidedly, Yes. But the fact remains, 'once a leper, always a leper.' . . . With regard to the vexed question of contagion, I am of opinion that leprosy is inoculable. . . . There is the fact that leprous sores do heal in a great many cases without any treatment whatever. . . . But these sores healing do not, as so many people imagine, mean that the leprosy is cured, but merely that a local manifestation has been cured; the leper remains a leper, and sores will certainly occur at some future period. . . . All the remedies seemed to be equally inefficient. Nerve remedies had, I presume, altered nerve tissue to deal with, therefore they could not act in their usual way. The much bepraised resorcin and icthyolin were as useless as all the others, and the conclusions I have come to are :—(1) Lepers should be well fed ; (2) kept scrupulously clean ; (3) have some inunction to keep the skin soft ; (4) have their sores treated on rational principles. Then their lives are made less burdensome to them, and they are comparatively happy. The remedies used were strychnine, phosphorus, arsenic, mercury, potassium iodide, chaulmoogra oil, resorcin and icthyolin, gurjun oil, neem oil, and strychnia, and sweet oil with chaulmoogra oil and gurjun oil internally. I made sketches of some of the patients I treated, with remarks made during the course of treatment. I treated in all fifty patients, with in no case any marked benefit as regards the cure of the disease."—*Calcutta Daily News, November 3, 1891.*

In an article on "Leprosy in Kashmir," by Ernest F. Neve, M.D., F.R.C.S., Ed., I find that, " as treatment, iodoform, iodide of potassium, mercury, arsenic, quinine, phosphate of soda, and mudar root, have been tried internally ; while balsam of Peru and gurjun oil have been rubbed on externally. Carbolic acid and iodine has been used for wounds. Nerve stretching has been adopted for anæsthesia and trophic ulcers. None of these is a specific."—*Lancet, 16th November, 1889, p. 1000.*

The *British Medical Journal,* August 10, 1889, contains an article, entitled, " Clinical Notes on Leprosy," by James J. L. Donnet, M.D., Inspector-General of Hospitals and Fleets, Honorary-Surgeon to the Queen. Under the head of "Treatment" is the following (p. 304):—"In the treatment of this disease the hygienic, the dietetic, and the palliative had more influence than the therapeutic. Where the cleansing of the skin by baths could be effected; where soothing or stimulating applications were made to ulcerated surfaces and to skin, to delay the distressing symptom of prurigo ; where good and abundant food was given, fresh air obtained, and exercise without fatigue, taken, with attention to full ventilation of inhabited rooms; where measures were adopted to afford recreation and gentle excitement, and thus divert the mind from the disease itself, a marked difference and a decided improvement, were the consequence. Under these influences the disease made little advance. But it is one that follows a determined course, rapid in some, more dilatory and seemingly stationary in others, but never retrogressing, always advancing."

"Where drugs are administered internally, I remarked that only those possessed of dietetic properties—as, for example, cod-liver oil—were of any value. Mercury, arsenic, iodide of potassium, assacú (obtained by incisions into the bark of the *hura braziliensis*), ammonia, and other preparations, each acted on the system, *modo suo*, but not in the measure or way hoped for."

Dr. Max Sandreczki, in an article entitled "A Study on Leprosy," in the *Lancet*, August 31, 1889, p. 424, says: —"As to the possibility of cure, one may say without fear of contradiction that among adults it is excluded." . . . . "The discovery of the bacillus has not hitherto advanced the curing of leprosy. Neither the transmissibility nor the mode of propagation has been demonstrated. Unfortunately, we are not permitted here to make dissections, and it is almost impossible to procure objects for the microscope. In conclusion, I would remark that if the bacilli cause leprosy, and propagate it from one person to another, how can we explain the long latent period—the repose of the bacillus for years? Is it not probable that the human body, more and more degenerated by years of misery and by every sort of hurtful influence, becomes the soil favourable for the development and culture of the bacillus?"

In a discussion at the Royal Medical and Chirurgical Society on the cure of leprosy, Mr. Macnamara, the author of " Leprosy a Communicable Disease," observed that "he had never seen a case which he could regard as being even relieved by treatment."—*The Lancet, May 31, 1890.*

Dr. Thin, under the head of "Treatment," p. 203 *et seq.*, observes that iodide of potassium has been experimented with by Danielssen on a large scale. It always produces a more or less violent eruption of nodules with feverish symptoms. Dr. Thin adds: "It affects patients both as a powerful poison and as a means of cure." After devoting several pages to recording experiments in drug medications by Dr. Unna, which must have caused much suffering to the hapless patients, Dr. Thin observes:—"Alas, that strenuous exertions directed with such intelligence and experience, should after all have turned out fruitless! The bacillus, however, remains apparently uninjured; and, although the treatment does not save the patient from his inevitable fate, Dr. Unna has done good service in making such an exhaustive experiment with strong drugs that had not been sufficiently tested previously."—*Leprosy, by Dr. George Thin, 1891, p. 214.*

Dr. Andres Navarro Torrens, Physician-in-Charge of the Provincial Hospital, Las Palmas, Canary Islands, writes to me, 1889:—"I have not up to this day seen any positive and evident cure by medicinal treatment."

A communication from the Superior Council of Health, Mexico, says:—"In the medicinal treatment there have been employed, successively and without result, mercurials, hydrocotila sciatica, guano, yodadurados, arsenic, sarsaparilla, and tarantula, as diaphoretic measures."

"The therapeutic as well as the dipterocarpic methods of treating the disease have been hitherto ineffectual."—*Leprosy in Foreign Countries, 1886, Honolulu, p. 186.*

Dr. K. Yamamoto, surgeon on board His Imperial Japanese Majesty's ship "Rinjio," in a communication to the Honolulu *Press*, June 19, 1883, describes leprosy as an incurable disease.

A communication to the Hawaiian Government from "the Faculty," Barcelona, Spain, in reply to an inquiry as to the results of medical treatment, says:—"A great number of medicines have been tried to combat this disease, but, in almost all cases, without result."—*Leprosy in Foreign Countries, Honolulu, 1886, p. 194.*

Dr. Kaurin, Medical Superintendent, Leper Asylum, Molde, Norway, says that "hitherto no specific remedy for leprosy has been found. At an early stage the disease may be cured by good diet and regimen; by careful nursing of the skin, baths, and symptomatic treatment."—*Journal of the Leprosy Investigation Committee, January, 1891.*

After much careful reading, and inquiring into alleged cures in various countries, including those of Dr.

Beauperthuy, of Venezuela; the remarkable results, reported by Colonel Chrystie and Father Muller in the Indian journals,[1] of Count Mattei's treatment; the cures officially reported by certain foreign medical experts in recent Government reports of Hawaii; Dr. Koch's tubercular inoculation; and the nerve stretching and ulcer perforation at the Mucurapo Asylum, Trinidad, I find no evidence to warrant the belief in their ability to cure this disease. In every case the alleged benefit is only transitory, the tubercles and abscesses reappearing, often with increased malignity.

TUBERCULIN IN LEPROSY.

Dr. P. Ferrari, in an article on "Koch's Tuberculin in Leprosy," says:—"Dr. Ferrari gives the conclusions of several observers who have experimented with the tuberculin in leprosy. Dr. Danielssen considers (1) that tuberculin in leprosy gives general and local reactions, the former generally coming on four to six hours after the first injection, but sometimes in twelve hours, and rarely in two to three days—the local reaction is more tardy; (2) that unfavourable consequences ensue to the patient, the disease being aggravated, and that the reactions have some similarity to those produced by the preparations of iodine in lepers; (3) that the lymph does not kill the bacilli, but seems instead to give them

---

[1] The latest communication I have seen from Father Aug-Muller, St Joseph's Leper Asylum, Mangalore, S.C., where the Mattei treatment has been tried, appears in the *Calcutta Daily News*, October 30, 1891, and the writer only claims "amelioration in the health of the inmates," but no case of cure. This is no more than the natural result of hygiene without medicine at all.

nutriment and favours their reproduction and circulation in the blood ; (4) that when immunity to the remedy is established the disease is in no way arrested, nor the bacilli destroyed. Dr. Ferrari has himself come to the conclusion, from the consideration of the above cases and of those of other observers, that tuberculin exhibits no direct useful action on the leper. As in tuberculosis, it may act on the torpid condition of the tissue, not so much by any specific effect as on account of the small resistance of the diseased tissue. He remarks particularly on the outburst of new tubercles during the paroxysms of fever."—*Journal of the Leprosy Investigation Committee, No. 4, December, 1891, pp. 46-47.*

Dr. J. L. Bidenkap, Physician to the Department for Skin Diseases, Rig's Hospital, and Lecturer on Dermatology at the University of Christiania, Norway, says, under the title of " Curative Treatment :"—" There has been searched in vain for remedies which have a direct favourable influence on the disease, or specifics. Mercurial preparations have, of course, been tried to a considerable extent. They have, however, but little or no influence on the disease, and often do harm. The reported few favourable results, or even recoveries, of the employment of these agents have possibly been due to accident, or in some cases to an erroneous diagnosis, the disease having been confounded with syphilis. The same is also the case with iodine preparations, which, according to the experience of Danielssen, are liable to evoke acute outbreaks, and on the whole are hardly advantageous, any more than bromine combinations. Antimonial preparations, and particularly tartar emetic, have, among others, been tried by Danielssen in increasing doses for long periods, but, it seems, without notable effect. Arsenic seems to have little or no influence, and this applies to most remedies of the same class. Of vegetable drugs there are a great many which have been vaunted as specifics at the places where the people understood to gather and to use them. In the European hospitals, however, these remedies have proved more or less inert. I have used salicylic acid for long periods of time; but, contrary to the experience of others, I have seen more harm than benefit from the use of it."

"Of therapeutic remedies it is mainly baths, tepid tub and steam baths, which seem to act favourably, and the general health of the patients often improves greatly by their methodic employment."—*Abstract of Lectures on Leprosy, 1886, pp. 65-67*.

In a Report on Leprosy in New South Wales, Mr. Edmund Sager, the Secretary to the Board of Health, Sydney, says :—" In adopting the system of segregating cases of leprosy, the Board has had before it the fact that the disease is, so far as at present known, incurable, and that its efforts must be directed to prevent its reproduction or spread."

The Report of the Inspector of Asylums, Cape of Good Hope, presented to both Houses of Parliament, 1891, referring to leprosy, p. 30, says :—" There is no reason to believe that specific treatment has in any case effected a cure. Gurjun oil, arsenic, potassium iodide, and icthyol have been tried without any result beyond what would be expected of healthy surroundings and good diet."

In the Report of the Select Committee on the Spread of Leprosy, Cape of Good Hope, under the head of "Minutes of Evidence," July 1889, I find the following :—

Dr. W. H. Ross, Police Surgeon at Cape Town for twenty-two years, under examination—

Q. 323. "During your stay at Robben Island has it ever been seriously attempted to treat leprosy as a disease in the same way as any other disease would be treated at a public hospital or asylum?"—"We endeavour to alleviate their sufferings, but very little can be done, except to treat symptoms and complications as they arise in this incurable and hopeless malady."

Dr. H. C. Wright testified—

Q. 21. "Leprosy is a disease for which as yet we have found no cure."

Q. 33. "Have you had any experience of the remedies tried for the cure of leprosy?"—"I do not know of any cure."

Q. 36. "You are, however, of opinion that leprosy is incurable?"—"I do not think that we have found any specific for it. I think all authorities are agreed as to that."

Dr. Beck testified—

Q. 176. "Leprosy cannot be cured, the disease must run its course, and the patient dies."

Dr. J. H. Cox testified—

Q. 227. "I believe that it (leprosy) is not curable."

The Report (page ix.) says:—" In the anæsthetic form of the disease there is not much actual physical pain, but in that as well as in the tubercular form the patient undergoes a gradual physical and often moral decay which renders him an object peculiarly deserving of the compassionate care of the State."—*Cape of Good Hope, Report of the Select Committee on the Spread of Leprosy, July, 1889.*

As to therapeutic means, Dr. Alexander Abercromby observes:—" The animal, the mineral, and the vegetable kingdoms may be said to have been ransacked. By some the chloride of mercury and the bichloride, in combination with sarsaparilla, have been highly extolled; whilst others, relying more upon antimony as a curative means, have given it as their opinion that mercury is positively injurious. In large doses all seem to agree that it aggravates the disease."—*Thesis on Tubercular Leprosy, p. 20.*

The same authority, writing April 20, 1892, from Cape Town, in a letter to myself, says:—" After more than thirty years experience as regards the treatment of the disease 'medicine makes no impression upon it even in the earlier stages.'"

In a Special Report on Leprosy from Robben Island for 1891, Dr. S. P. Impey, Medical Superintendent, says:—" A good deal is said and done when small-pox threatens our country, but how much more dreadful is leprosy than small-pox. Small-pox may kill hundreds in a few days, but all who suffer do not die. Leprosy is fatal in all cases, and all who once catch it, after lingering for a few years a life worse than death, shunned by all, they become outcasts, with no hope of cure before them, and die objects of abhorrence and pity."—*Reports presented to both Houses of Parliament by Command of His Excellency the Governor of the Cape of Good Hope, Cape Town.*

The *Lancet*, of April 16, 1892, has a leading article on the results of this mode of treatment of leprosy. Referring to the experiments of Dr. Danielssen (of Norway), it says:—

"The injections were made daily, unless the reactions were severe, when an interval of several days was allowed. In some of the cases the treatment had to be stopped even when only small doses had been reached, because the eruptions in the tuberculated and in the anæsthetic cases became so intense that the disease was evidently aggravated." . . . Drs. Babes and Kalindero, of Bucharest, treated seven patients with inoculations of the same material. Upon these the *Lancet* observes: "Unfortunately the conclusion drawn was that tuberculin aggravated the disease considerably, and, by setting free the bacilli, started fresh *foci* of the disease, and made the whole process more active. As in lupus and phthisis, the patients became tolerant of the tuberculin after a time; but the disease progressed all the same, and fresh symptoms were frequently excited; many also of the old lesions became red and sensitive. In the anæsthetic form the patches enlarged, became redder and more sensitive, and new patches appeared." . . . Dr. Colcott Fox's patient, treated by Mr. Cheyne at King's College Hospital, seems to have fared no better. The injections (of tuberculin) "were followed by severe pains in the ulnar nerves, and in the calves and knees, with a temperature of 103° F. After the fourth injection extreme and dangerous collapse followed the reaction, the eruption became more prominent and large blebs formed on both feet, and there were other unpleasant symptoms. All these symptoms occurred in a purely non-tuberculated case of long standing which was almost quiescent, obscure pains at intervals being the only symptoms left; the injections therefore had obviously excited a neuritis and revived the typical macular eruption which had faded away." . . . In one case (treated by Dr. Goldschmidt, of Madeira), "intense and prolonged reaction ensued, followed by blebs and new swellings of skin and mucous membrane, while on the lower limbs absorption of the leprous infiltration was produced." . . . "Mr. Cantlie of Hong-Kong used it (tuberculin) in seven cases, and he also found bacilli in the blood after injection, there having been none there before." . . . "In Dr. Radcliffe Crocker's case a single injection of two milligrammes excited an attack of leprous fever, which lasted three weeks. Scores of new tubercles came out all over the body, but under gurjun oil inunctions they all disappeared again, and there was a little less infiltration in some parts. The patient, a tuberculated leper, had previously improved considerably with large doses of chaulmoogra oil, and had had no ebrile exacerbation for two years prior to the injection. As an ultimate result he was not really any worse; but three weeks' fever was a long price to pay for a single injection, and he did not care to have it repeated."

"The above cases do not exhaust the list of experiments, but they are sufficient to show that tuberculin is very uncertain in its immediate effects on leprosy; that while in some it produces no reaction or effect at all, in others, even with small doses, considerable and prolonged reaction may ensue; and that therefore not more than one milligramme should be given at first, and cautiously increased according to the patient's toleration of the fluid. Secondly, that it revives the activity of cases which have been quiescent for a long time, producing neuritis, bringing fading rashes into prominence, exciting bullæ, and setting free the bacilli; these may get into the circulation and produce fresh cutaneous and other lesions. In the ultimate effect, while some patients have shown improvement that has not been proved to be permanent, and in some was certainly only temporary, on the whole the position may be summed up by stating that, as far as the results hitherto obtained are concerned, the improvement which may result is too uncertain, too limited in character and time, and purchased at too great a risk of aggravation of the disease by the dissemination of new foci, for it to be recommended as a treatment for leprosy; and although some of Dr. Hunter's or similar modifications of tuberculin might lessen the immediate disagreeable effects, the fact of its action being to set free bacilli rather than to destroy them should make us seek in another direction for remedial agents for the relief of victims of this much-dreaded disease."

It is admitted that lepers suffer acutely from neuralgic pains, which Dr. Sidney Bourne Swift says "are common amongst lepers of all types;" and the wretchedness of their condition is intensified by the drastic treatment adopted at many of the hospitals and lazarettos. The application of Koch's tuberculin is often attended with excruciating pains.

Under the head of "Cantharidin Treatment in Leprosy," the *British Medical Journal*, September 5, 1891, says:—"The injections caused severe pain, but no local reaction; they were, however, always followed by a rise of temperature in the leprosy cases."

The writer, when visiting a leper hospital a short time ago, was witness of the fashionable inoculative experimental treatment. The poor creatures were brought into the surgery one after another, some brave and others with a timid, appealing look in their eyes. To enable them to bear the pain of the hypodermic syringe, thrust by the operating physician deep into the flesh, they had a handkerchief between the teeth while held by the hospital nurse or attendant. The puncture of the instrument is usually the least painful part of the experimental process. The treatment, which is often continued for months, produces sickness, acute headaches, and fever.

The rage for experimental research has long since passed the bounds of decent humanity, and many who have investigated the facts are of opinion that legislation ought to be specially invoked in the interest of these, the most hapless members of the human family. It is a common experience for lepers in hospitals to attribute the aggravation of their maladies to therapeutic treatment in the hospitals, and this is confirmed by high medical authority.

In the report on "Leprosy and Yaws in the West Indies," by Dr. Gavin Milroy, London, 1873 (c. 729), the author observes :—" It struck me forcibly, on observing the persistency of the anæmic condition of so many patients at Kaow Island (the Leper Settlement of British Guiana), and still more of Dr. Beauperthuy's three private patients at Bartica, that this symptom was in part due to the prolonged use of a medicine, which is found to be notably injurious in like conditions of the system in European practice." From the foregoing testimonials (and others which I have not space to adduce) confirmed by my own observations and inquiries, it appears to be the general experience that leprosy attacks most readily those whose vitality is reduced by malarial fever, syphilis or insanitary conditions—*i.e.*, unwholesome food, and impure water, and is most speedily fatal to those unfortunates at the hospitals who have been selected as subjects for inoculation experiments, mercurial treatment, and other drug medication. Whatever benefit patients have obtained is admitted, by the highest authorities in all the countries I have visited, to have been due solely to improved sanitary conditions and hygienic treatment. One thing is certain, the unfortunate patients dread the experimental treatment to which they are subjected by lepra experts, often escaping from lazarettos and secreting themselves in the gullies and fastnesses of the hilly regions and in the jungle to avoid the terrible ordeal to which they would probably be subjected.[1]

It is acknowledged that there is no specific for leprosy, nor is there any drug which has permanently ameliorated the condition of the patient. Where gurjun, chaulmoogra, or other oils have been used externally there is a temporary alleviation probably due to the massage or friction employed, but the disease, when once it has taken hold of the system, is absolutely unamenable to therapeutic treatment. On the other hand, there appears to be a consensus of opinion that, in the majority of cases, the condition of the lepers may be improved, and life rendered more tolerable, by their removal to a salubrious locality, with wholesome food, cleanly and orderly habits, cheerful recreation and employment for body and mind, and that under these conditions the disease may in rare instances die out. The worst picture of concentrated and hopeless misery I have ever seen was in 1884 in the Lazaretto of Damascus, where all these essentials were wanting. In view of these experiences, which can

---

[1] Dr. George L. Fitch, formerly Medical Superintendent, Leper Settlement, Kalawao, Hawaii, says:—"On November 14, 1883, I inoculated six lepers with the virus of syphilis, by taking six ivory vaccine points, and scraping off the surface of a mucous patch on the inner side of the lower lip of a native woman. The points were then allowed to dry, and three hours afterwards I transferred the virus to the arms of six leper girls under twelve years of age. December 14 following I repeated the experiment, taking the virus from a hard chancre on a Portuguese who came to my office. I saw this man in March, 1884, three months later, in the office of a brother physician, and found he was suffering from secondary syphilis. The last time I used fourteen points and inoculated fourteen lepers therefrom, but no result followed in any of the twenty experiments."—*New York Medical Record, September 10, 1892, p. 297.*

be multiplied to any extent, and on the ground of humanity, is it not time to put a stop to the torture to which the incurable sick lepers are subjected by drug medication and inoculation, and let these miserable creatures be made as comfortable as tender nursing, varied occupations, and amusements will allow, and permit them to die in peace? Not a few of them are the victims of the Jennerian system, and these are the smallest compensations we can make for the irreparable injury done to them.

Above all, we can cease to propagate this fell disease. We can discontinue the enforcement of a medical practice which experience has shown to be a potent factor in its dissemination. This is the most terrible count in the long indictment against vaccination; but, as I have shown, it is amply sustained by the testimony, often unwilling testimony, of unimpeachable witnesses of the highest credit and authority. It may perhaps be thought that what I have quoted is only *ex parte* evidence, that possibly an equal array of high authorities may be cited on the other side. This, however, is not the case. There is here no conflict of testimony, so far at least as inoculation is concerned, and vaccination is one and the chief means by which inoculation is effected. If there are authorities of equal weight to be thrown into the opposite scale, in my careful personal inquiries and patient investigation into the subject I have not discovered them.

SUMMARY.

The results arrived at in this volume may be briefly summarised as follows :—

(1.) That leprosy has greatly increased during the last half century, and that it is prevalent in many places where it was formerly unknown.

(2.) That whilst the opinion of medical authorities and experts varies considerably on the subject of the contagiousness of leprosy, the preponderance of authority is in favour of the theory that it is not contagious in the ordinary sense of the term, but is communicable by means of a cut, sore, or abraded surface ; and this view is confirmed by my own personal investigations.

(3.) That other alleged factors such as malaria, a fish diet, syphilitic cachexia, heredity, and insanitation are admittedly unequal to explain the rapid growth of the disease in certain of our crown colonies and dependencies, as well as in other countries.

(4.) That on one point there is much agreement and hardly any dissent, namely, the inoculability of leprosy ; and that the view of leprosy as an inoculable disease, while it is most clear to those who take the malady to be due to a bacillus, is older than the bacteriological evidence, and is not dependent thereon.

(5.) That the most frequent opportunities of inoculating the virus of leprosy are afforded in the practice of vaccine inoculation, which is the only inoculation that is habitual and imposed by law; and that the evidence here adduced is calculated to show that vaccination is a true cause of the diffusion of leprosy.

(6.) That the official information, collected by interrogatories and otherwise, has not been hitherto of a kind to show how far vaccination has determined the amount of leprosy in recent times; and that any interrogatories that may be sent out in future should not be limited to ascertaining the effects, as regards leprosy, of hypothetically "pure" lymph. When on very rare occasions interrogatories have been submitted, they have been framed to ascertain the results of a purely hypothetical system of vaccination which is not anywhere discoverable in practice (*i.e.*, with pure lymph, and free from hereditary taint), and the replies are therefore futile and misleading.

(7.) That with the exception of two groups of cases—those adduced by Dr. Roger S. Chew, of Calcutta, and Dr. S. P. Impey, of Robben Island—those reported in this volume have not been the result of special investigation, but have cropped up accidentally in the course of medical practice, and in some instances have been published by practitioners with apologies to the profession for presenting such unwelcome disclosures.

(8.) That the increase of leprosy in the Sandwich Islands, the West Indies, the United States of Colombia, British Guiana, South Africa, and New Caledonia, has followed *pari passu* with the introduction and extension of vaccination, which in nearly all these places, without previous inquiry or demand from the inhabitants, has been made compulsory.

(9.) That as leprosy is a disease of slow incubation, often taking years to declare itself, and in its incipient stages can be detected only by practitioners of large experience, it follows that, in countries where leprosy exists, there is great danger of extending the disease by arm-to-arm vaccination.

(10.) Leprosy being one of the most loathsome diseases to which the human race is subject, and being practically incurable, it behoves all interested in the public well-being to do their best to *prevent* its diffusion, and, as a means thereto, to discourage the practice of vaccination on that ground, if on no other.

APPENDIX.

THE following items relate only incidentally to the main object of this volume, but may be useful to the reader as illustrating the methods by which vaccination has been fostered and made obligatory in some of our possessions.

These appendices refer briefly to the mischievous consequences of vaccination, and demonstrate its failure from its inception either to mitigate or to prevent small-pox, with facts showing the growing opposition to this form of State medicine, and the necessity of substituting sanitary amelioration.

Some particulars are also furnished of a medical vaccination census carried out in the year 1883 by a committee of vaccine experts, with the object of reinstating in public favour a practice which had been discredited by numerous vaccination fatalities, and notably by the Misterton, Sudbury, and Norwich vaccine disasters. Particulars of the more recent disastrous results of vaccination will be found in the third report of the Royal Commission on Vaccination. London: Eyre & Spottiswoode, 1890 (c. 6192).

VACCINATION FRAUDS.

The following is taken from the "Life of Jenner," by Baron, a warm partisan of vaccination, published in 1827, vol. i., pp. 557-559:—

"On the introduction of vaccine inoculation into India, it was found that the practice was much opposed by the natives. In order to overcome their prejudices, the late Mr. Ellis, of Madras, who was well versed in Sanscrit literature, actually composed a short poem in that language on the subject of vaccination. This poem was inscribed on old paper, and said to have been *found*, that the impression of its antiquity might assist the effect intended to be produced on the minds of the Brahmins while tracing the preventive to their sacred cow. The late Dr. Anderson, of Madras, adopted the very same expedient in order to deceive the Hindoos into a belief that vaccination was an ancient practice of their own. . . .

"Shortly after the introduction of vaccination into Bengal, similar attempts were made to prove that the practice was previously known there also. . . . A native physician of Bareilly put into the hands of Mr. Gillman, who was surgeon at that station, some leaves purporting to contain an extract of a Sanscrit work on medicine. This work is said to be entitled Sud'ha Sangreha, written by a physician named Mahadeva, under the patronage of Rájá Rájusin'ha. It contained a chapter on Masúrica or Chicken-pock. Towards the close, the author appears to have introduced other topics; and immediately after directing leeches to be applied to relieve bad sores he proceeds thus : 'Taking the matter of pustules, which are naturally produced on the teats of cows, carefully preserve it, and, before the breaking out of small-pox, make with a fine instrument a small puncture (like that made by a gnat) in a child's limb, and introduce into the blood as much of that matter as is measured by a quarter of a ratti. Thus the wise physician renders the child secure from the eruption of the small-pox.' This communication was shown to Mr. Colebrooke and Mr. Blaquiere, both eminent Sanscrit scholars, and they both suspected that it was an interpolation. . . . I believe I may further add that Mr. Colebroke made inquiries whilst in India, which fully satisfied him that no original work of the kind ever had existence. Sir John Malcolm has also been kind enough to ascertain that no such book is to be found in the library of the East India Company. From these statements it must be apparent, that the well-meant devices of those who attempted to propagate vaccination in India have led to the belief that the practice was known to the Hindoos in earlier times."

It may be added that Dr. Anderson, above referred to, is congratulated by Jenner's biographer on his "unceasing exertions at Fort St. George."

COMPULSORY VACCINATION IN BOMBAY.
THE DANGER OF TUBERCULOSIS.

Notwithstanding the proofs laid before the Royal Commission on Vaccination regarding the futility of the practice as a prophylactic against small-pox, its injurious consequences in spreading inoculable diseases, and the cruelty and injustice attending its enforcement, there are not wanting those who are continually plotting to extend the system, by means of coercive legislation, amongst populations who are known to entertain a widespread repugnance to vaccination, but who are without representative institutions. At a meeting of the Bombay Legislative Council (reported in the *Times of India*, February 24, 1892) held for the purpose of hearing the second reading of the bill to prohibit the practice of inoculation, and to make vaccination compulsory in certain districts of the Bombay Presidency, the Honourable Mr. Javerilal U. Yajnick moved certain amendments to the law. In the course of his argument, in which he points out the danger of transmitting leprosy and syphilis by means of arm-to-arm vaccination, he quotes the opinion of "an able and experienced medical gentleman," Dr. Bahadurjee, who in reply to an inquiry writes :—

"In answer to your letter in which you ask me my personal opinion on the arm-to-arm vaccination method which is intended to be enforced by the new Vaccination Bill, I have no hesitation in saying that, besides its being not suited to the peculiar conditions which obtain in this country, on professional grounds the method is objectionable, and for these reasons :—
1. Arm-to-arm vaccination obviously acts as a channel for the transference of some skin diseases, and affords a ready means for propagating such inherited constitutional taints as those of syphilis and leprosy. No doubt, special rules, with full details, will be framed for the guidance of the operators in their selection of proper subjects, with a view to avoid these

mishaps; but, having regard to the class of men from whom the supply of district vaccinators is to be obtained, the detailed rules will be of as much use to them as the paper on which they are printed. 2. Syphilitic taint does not necessarily show itself in ill-health at the early age at which vaccination is practised and demanded by law. A child may be in fair health, and yet have inherited syphilis. Moreover, syphilis does not stamp itself on the face and arms, so much as on the back and legs—parts not generally examined by the vaccinator, and thus apt to be overlooked Only yesterday I was asked to see a case of skin disease in a child. On stripping the child bare, I found him fairly healthy to look at, and could see no skin blemish on his person. But closer examination of the hidden parts revealed the presence of unmistakable condylomata (syphilitic). These condylomata unnoticed, I should have passed the child as a very fair specimen of average health, and a fit subject to take the lymph from. Syphilis, as betrayed in obtrusive signs, is not difficult to recognise, but when concealed, as is more often the case, it is by no means easy to detect it. 3. In the case of leprosy it is still worse. There is no such thing as a leper child or infant. The leper heir does not put on its inherited exterior till youth is reached. And it is by no means possible by any close observation or examination of a child to say that it is free from the leprous taint. Surely arm-to-arm vaccination will not help to stamp out leprosy. On the contrary, it has been asserted, and not without good reasons, that it has favoured the propagation of the hideous disease. 4. It is acknowledged that extreme care is required in taking out lymph from the vesicles to avoid drawing any blood, for blood contains the germs of disease. Extreme care means great delicacy of manipulation, and delicacy of manipulation with children is not an easy task, and requires some experience and training. Is this to be expected from the class of men who are going to act as public vaccinators in the districts? Supposing a district vaccinator to acquire it to some extent after considerable practice, what about the delicacy of manipulation of one newly put on? 5. Puncturing a vesicle with such delicacy as not to wound its floor and draw blood is one great difficulty. But the selection of a 'proper' vesicle is another as great if not a greater difficulty. Products of inflammation are charged with the germs of disease, the contagion or contamination media, as much as the blood itself is. And the contents of an inflamed vesicle are quite as contaminating as the blood itself of a subject who, though charged with the poison of (inherited) syphilis or leprosy, has none of the obtrusive signs of the taint for identification. And out here inflamed, *i.e.*, angry-looking vesicles are not the exception but the rule, as can be easily told by personal observation and experience, and equally easily surmised if the habits of our poor be duly considered. Thus, even if no blood be drawn, the danger of transferring constitutional taints by the arm-to-arm method is by no means small ; remembering that leprosy which claims India, and not England, for one of its homes, does not admit of any detection on the person of a subject from whose arm lymph may be taken, and that syphilis is more often difficult to detect than otherwise, and remembering, also, that both these are often met with largely in some districts."

With regard to tuberculosis, the most deadly of all diseases in Europe, the following extracts from the translation of an article in the *Gazette Hebdomadaire des Sciences Médicales*, by Dr. Perron, *Officier de la Légion d'Honneur*, which appeared in the *Vaccination Inquirer and Health Review* of December, 1890, may arrest the attention of the reflective reader. Dr. Perron says :—

"The possibility of conveying tuberculosis to man in the act of vaccination was long ago pointed out. Tuberculosis has, in fact, a special predilection for the bovine race which yields us our vaccine. There are few of these animals that escape its attacks ; the calf, the heifer sometimes bear traces of it but a few weeks after their birth. It would then appear quite natural to suppose that the vaccine, taken from a bovine animal and inoculated by the skin, might thus convey tuberculosis to the vaccinated subject. It is by no means so, however ; for it is demonstrated that the inoculation of tuberculosis by way of the skin is extremely difficult in itself, and that there is not the slightest fear of doing so by way of the vaccinal punctures. The direct conveyance of the tuberculosis contagion in this manner need not, then, be taken into account. If vaccination renders man more prone to contract tuberculosis, it is, in our opinion, by a method altogether different.

"We hold that we must, in this case, arrange our facts in accordance with the new theories of which we have spoken above ; that we must, that is to say, consider, with respect to vaccination and the possibility of a tuberculous contagion due to it, the part which can be played in the organism by nocivity, or receptivity, in relation to micro-organisms.

"The cow, as we have said, is the tuberculous animal *par excellence* She is often the bearer of specific granulations, sometimes even along with the appearance of ordinary health. She is, therefore, a soil eminently favourable, and therefore very receptive, for the bacillus of Koch. But along with tuberculosis there is another acute malady specially attaching to bovines, for it possesses the property of arising spontaneously among them, namely, the vaccinal disease, which, as we all know, shows itself locally by the appearance, on the teats and on the udder, of pustules, whereof we avail ourselves for human vaccination. Thus, then, two acute diseases, tuberculosis and vaccinia, find always in the cow the soil most favourable for their evolution; and that clearly because the medium of cultivation is propitious both to the bacillus of Koch and to the micrococcus of vaccine.

"If, as announced by Professor Bouchard, the medium created by a vaccination can be destructive to one or several microbic species, we may add that, by the law of reciprocity, a medium of cultivation may at the same time be favourable to one or several microbes. That is exactly what happens with the cow in respect of tuberculosis and vaccinia, diseases between which the soil of cultivation establishes, as we see, a striking connection.

"This is the time to examine what happens when we inoculate a human subject with cow-vaccine. By that act we bring the human organism into a state of immunity, which is certainly bactericidal as against the microbe of variola; that is the benefit we seek, and which constitutes the vaccinal immunity. But here is the important point; at the very time when we have created in the man the vaccinal soil, we run the risk of having, *ipso facto*, established that humoral state (*terrain humoral*) which is favourable to the tuberculous genesis, that is, the medium of culture which is receptive for the bacillus of Koch.

"The first and the most grave result which follows from this interpretation is, then, that vaccination, besides the advantages which it offers us in our contest with variola, presents the danger of opening the way for the invasion of tubercle.

"If we now turn back and examine the events of the last century or so, we can show a constant increase of tuberculosis, a fact never hitherto satisfactorily explained. There was a time when this malady existed only as an exceptional thing; now, actually, in spite of the incessant progress in public and private hygiene, in spite of all the material improvements that have been made, it tends more and more to rise to the rank of a pestilence. It should be remarked that it strikes by preference at the young lives, that is, those who are, nevertheless, at ·the age when the physical resistance to morbific causes is the strongest. Now, a malady which originates in exhaustion, in vital poverty, should display its power in the inverse order, and should fall most heavily on the old. We are, then, compelled to believe that young folk offer, for some quite special reason, an exceptionally favourable soil for the implanting of Koch's bacillus.

"Side by side with this growing extension of tuberculosis, we see developing, *pari passu*, and in the same period of time, that is to say, since the beginning of the century, the practice of vaccination. We may ask ourselves whether in this double simultaneous evolution there is not a hidden oneness? If tuberculosis, in spite of all sanitary precautions, has multiplied its attacks during the last hundred years, it is, we submit, because vaccination has come to create for it a propitious soil. That would explain, not only its advancing growth in all civilised countries, but also its special influence over the young subjects who are always more or less recently vaccinated, and consequently more receptive than the others in the presence of the bacillus.

"In all European armies, vaccination is the order of the day. On their arrival with their corps, the young soldiers are forthwith carefully re-vaccinated. Now, the military statistics of all countries show an enormous proportion of various forms of tuberculosis among soldiers, especially during the first and second year after their enlistment. Divers causes have been invoked for the explanation of the facts. First, the moral decline produced amongst young soldiers by their separation from their families. That might have been possible formerly, but it is not probable in our day when facility of communications generally permits the young soldier to remain in touch with his native country. Besides, in the army afloat, which is less favoured in this respect, we find no more cases of phthisis than in the army ashore. Nor can a bad hygiene be any longer pleaded, nor an inferior dietary, for the European States take the greatest pains to secure for their soldiers the best of material conditions, and succeed in doing so to a very satisfactory extent. Nor can over-work be alleged, for in time of peace the routine of the service requires, save under very exceptional circumstances, just as much exercise as goes to make up a healthy amount of daily exercise. To sum up, the young soldiers find with their corps material conditions of life, which, for a very large number, are superior to those of their native surroundings. Their life in the great towns, though evidently having an injurious effect, cannot by itself explain the numerous cases of tuberculosis of which we are speaking, for the barracks are in general well situated and looked after in accordance with the rules of health. Whence then can come these attacks of tuberculosis, so sudden, so numerous, upon subjects that, but a few months before, the council of revision rightly declared to be fit for service. Tuberculosis of the lungs, of the organs, of the joints, of the bones, etc., all these fatal evils show themselves in the garrisons of all countries with a frequency before which one might well despair. We believe that we must simply seek the reason for these facts in the re-vaccination which awaits the recruits upon their arrival at their corps, and which transforms them forthwith into a medium which is receptive towards those germs of tubercle which swarm in centres of population. This re-vaccination immediately upon enlistment is all the more regrettable and inopportune since just at that moment the young man, separated from his family, his country, his familiar conditions of life, undergoes, without any period of transition, total and radical changes in his manner of life, and thereby finds himself less well equipped for resistance."

Referring to the efforts made by the Indian health authorities to escape the dangers of inoculated diseases by the introduction of animal lymph, the Calcutta *Daily News* of 9th February, 1892, says:—" In trying to avoid the Scylla of leprosy, syphilis, and kindred evils, they fall into the Charybdis of tuberculosis and other equally fatal maladies. When doctors disagree, patients usually have the reverse of a comfortable time, and in the present situation the public may well look around in alarm, and cry to be saved from the dangers with which the whole subject of vaccination seems to be beset."

THE REVOLT AGAINST COMPULSORY VACCINATION IN INDIA.

The first triennial report of the working of the vaccination department in Bengal has been recently published, and the Commissioner says that the Acts in the rural districts are practically a dead letter; vaccination is rejected by all the higher class of Hindus—the Brahmins, Marwaries, Rajputs, and Burmahs—while, among the Mahommedans, the Ferazis display the utmost repugnance to the Jennerian rite. In nearly every village, reports the Commissioner, there are families who persistently refuse vaccination, and secrete their offspring to escape the vaccinators. The *Madras Mail* of July 2nd, 1890, says that vaccination is very unpopular with many classes; and the *Madras Times*, April 16th, 1891, reports that summonses were issued against fifty-two recalcitrants for non-vaccination. At Midnapur, Bengal, where the vaccinations formerly averaged three thousand annually, the number fell to nine hundred last year.

The *Allahabad Pioneer*, September 23rd, 1891, says :—

"The Civil Surgeon of Coconada, in his report on vaccination in that town, says that it is a common occurrence for parents to wash out the vaccine virus immediately after vaccination ; and the vaccinators further assured him that the natives are in the habit of rubbing in chalk, chunam, or flour, with a view, if possible, of preventing the vesicles rising on their children's arms."

The *Times of India*, July 14th, 1892, says :—

"The prejudice against vaccination in Burmah seems to be growing to quite a remarkable extent. The report for 1891-92 shows that in Lower Burmah the number of cases was only 129,509, or 10,812 less than in the previous year. In one district alone, Henzada, there has been a fall in two years from 10,134 to 5180, while in the Toungoo district the figures have declined from 8905 to 3069. The Prome, Thongwa, and Thayetmyo districts are also among those which exhibit a considerable decrease, and in most cases no explanation of the decline seems to be forthcoming ; while such explanations as are offered the officiating Chief Commissioner 'cannot regard with any satisfaction.' In Upper Burmah there was an increase of some 20,000 cases, but this seems mainly due to the extension of the Act. In Upper Chindwin there was a great and unexplained decrease, and five other districts also show a decline, while in some of these and a number of other districts the people put every possible obstacle in the way of vaccination. *At Katha, Mohayon, and Mobin so strong is the prejudice against arm-to-arm vaccination that the vaccinators appear to have narrowly escaped violence at the hands of the villagers, who organised an open resistance to the system.*"

The *Civil and Military Gazette*, Lahore, August 8th, 1892, says :—

"It appears that the natives of Lahore are opposed to compulsory vaccination. The inhabitants of several *mohallas* in the city have drawn up memorials to the local Government asking that the resolution of the Municipal Committee for the introduction of compulsory vaccination in Lahore be cancelled."

The *Allahabad Pioneer Mail*, 6th October, 1892, under the head of Vaccination in Bengal, says :—

"There has been a falling off of nearly two hundred thousand, or about 11 per cent. in the number of vaccination operations performed during the past year in Bengal, as compared with the record for 1890-91. The number of operators has meanwhile increased by nearly one hundred; and an analysis shows an average decrease of about 107 cases per operator."

And in a leading article the editor observes :—

"If anti-vaccinationists can be numbered in their thousands in England, it is small wonder that they can be numbered in their millions in India."

Those familiar with the social condition of India are aware that every effort has been made to remove this dread of the operation which exists more or less all over the country. New lancets and scarifiers have been introduced, and various viruses have been experimented with, one after another—cow, calf, sheep, goat, lamb, buffalo, and donkey lymph — the last, the discovery of Surgeon O'Hara, having been specially urged upon the attention of District Boards and municipalities by the Government. Surgeon-Major W. G. King writes to the Indian *Medical Record* that he is using *vesicle pulp* or "lanoline vaccine," which is applied by stretching the scarifications and "alternately dabbing and rubbing in the paste." Buffaloes, he observes, appear likely to yield very much more vesicle pulp than calves, but they exhale an "abominable odour," which renders the work of collecting the pulp most repulsive. The Commissioners state that only the lowest and most ignorant classes readily submit. The laws enforcing vaccination in British India, which are unparalleled for their severity, were passed without the consent and against the wishes of the people, whose objection to vaccination arises from a knowledge often gained by sad and bitter experience. They know that the fearful spread of leprosy in India and other countries is coincident with and, as they believe, due to the extension of vaccination, and they prefer to face the severities of the law, with its ruinous judicial penalties, or even to risk the dangers of the jungle, where they are sometimes compelled to seek refuge for their little ones, to the risks of this hideous and destructive scourge. That leprosy, confessed to be incurable, is inoculated by vaccination (a fact once vehemently denied) is now reluctantly admitted by the leading dermatologists of all countries, and by the most experienced chiefs of the leper asylums and public health departments in the West Indies, in South America, South Africa, and in the Sandwich Islands.

The *Madras Times*, May 18th, 1892, says :—

"Every effort is probably made to obtain pure and healthy lymph, but if the causation of leprosy has not yet been satisfactorily traced, no guarantee can be provided against the presence of the germs of the disease in the lymph used for purposes of vaccination."

VACCINATION IN THE WEST INDIES.

BARBADOS.

There is no compulsory vaccination in this island. The editor of one of the leading journals writes, 2nd May, 1890, that the feeling in the island is so strong against vaccination, that the advocates of vaccination are afraid to move in the matter, and any attempt to enforce it would probably create a riot.

In conversation with all classes and conditions of men (from December to February, 1888-9), from the Chief Justice, Sir Conrad Reeves, to the poorest boatman or sugar plantation labourer, from one end of the island to the other, I failed to discover a single advocate of compulsion. "Let those have it who want it, but don't force it upon me and mine," was the general reply. The police and postmen also get along admirably without re-vaccination. Epidemics are considered less frequent than in other well vaccinated districts, notably Jamaica, Martinique, Guadaloupe, and Hayti.

The natives are a proud, independent, and more intelligent coloured population than any I met with in the West Indies, and rather look down upon other islanders who are without political representation, and are subject to enforced vaccination. The population in April, 1891, was 182,206, or 1096 to the square mile—one of the densest in the world.

GRENADA.

In marked contrast to the parental freedom enjoyed by the inhabitants of Barbados, with its popular constitution and representative government, is the position of Grenada. I copy the following from the *Grenada People*, June 9th, 1892, concerning the oppressive legislation in one of the most beautiful islands in our West Indian possessions, administered as a Crown Colony:—

"During this week, upwards of thirty or forty of the peasants have been hauled before the police magistrate of the Southern District for alleged violation of the Vaccination Act. In nearly every case fines of half-a-crown have been imposed, representing almost half of the week's wages which these unfortunates, if they are employed, can hope to earn. In face of the Royal Commission on Vaccination, we do not see why the law making vaccination compulsory should be still enforced. At most, it is of doubtful benefit; and doctors differ as to the positive good or injury which it does. The advocates of Jenner's specific can quote very few cases, if any, in its support; whilst its opponents point with force and truth to the positive injury it has inflicted. Here, in Grenada, pure lymph is seldom employed. As a consequence, many of the children submitted to the process of vaccination contract therefrom fatal diseases. The lymph, in many cases, is collected from children inheriting a taint of the scrofulous disease which prevails amongst the peasantry; and many an otherwise healthy child, after the process of vaccination, presents the appearance of a disgustingly yawsey patient.[1] As eminent medical men differ as to the value and utility of vaccination, we think it ought not to be made an offence punishable by fine or imprisonment if parents refuse to vaccinate their children; but that the law should be amended in the direction suggested by the Royal Commission in their recent report, *i.e.*, it should be optional with the parent whether the child should be vaccinated or not."

VACCINATION A FAILURE AND A DANGER TO HEALTH FROM ITS INCEPTION.

Early in the century, when cases of injury were first charged to vaccination by either the suffering victims or their relatives, they were met by emphatic denials on the part of the supporters of Jenner. Proofs of the failures and mischievous results of vaccination, as shown by Dr. Creighton in his remarkable historical work, "Jenner and Vaccination," accumulated from all quarters, but still the vaccinators held on to their creed, Jenner having supplied them

---

[1] "Yaws" is a loathsome skin disease allied to syphilis.

with a theory broad enough to meet any contingency. He says there were some varieties of spontaneous eruption, all of which produced sores on the milkers, but only one of these was the *true* cow-pox, all the others being spurious and exerting no specific protective power over the constitution.

On the 18th June, 1890, evidence was adduced by me before the Royal Commission concerning the early failures of vaccinators and. the mischievous effects of vaccination.[1] In the year 1806 the College of Surgeons instituted an inquiry into the results of vaccination. The results of this inquiry have been overlooked by all the advocates of compulsory vaccination, and notably omitted from Sir John Simon's remarkable "Papers relating to the History and Practice of Vaccination, presented to both Houses of Parliament," London, 1857.[2] This report states that on the 15th December, 1806, a circular was drafted and referred to the Board of Curators, and, having received their approval, was despatched to 1100 members of the College in the United Kingdom, submitting the following questions:—

"1st. How many persons have you vaccinated?

"2nd. Have any of your patients had small-pox after vaccination?

"3rd. Have any bad effects occurred in your experience in consequence of vaccination? and, if so, what were they?

"4th. Is the practice of vaccination increasing or decreasing in your neighbourhood? if increasing, to what cause do you impute it?"

To the 1100 circulars only 426 replies were received. Why nearly two-thirds of the members kept silent, when at the outset they were converted in multitude to vaccination, was left unexplained. The replies were thus summarised by the Board on 17th March, 1807—

"The number of persons stated in such letters to have been vaccinated is 164,381.

---

[1] See Third Report of the Royal Commission on Vaccination, pp. 150-151.

[2] The report of the Royal College of Physicians, which was favourable to vaccination, was included by Sir John Simon in these "Papers." The corresponding report of the College of Surgeons, which appeared in the same publication, but was adverse to vaccination, was ignored — an important historical omission.

"The number of cases in which small-pox had followed vaccination is 56.

"The Board think it proper to remark under this head that, in the enumeration of cases in which small-pox has succeeded vaccination, they have included none but those in which the subject was vaccinated by the surgeon reporting the facts.

"The bad consequences which have arisen from vaccination are—

> 66 cases of eruption of the skin,
> 24 of inflammation of the arm, whereof
> 3 proved fatal."

A copy of the original report containing these remarkable admissions was produced by me before the Royal Commission, and examined by the president and each member of the Commission then present.

I have before me a copy of Volume VIII. of the *Medical Observer*, an ably conducted Metropolitan journal published in 1810 (produced also before the Royal Vaccination Commission), and on pp. 183 to 197 I find recorded (with chapter and verse for reference to the authorities) the particulars of 535 cases of persons having small-pox after vaccination; also similar details of 97 fatal cases of small-pox after vaccination and of 150 cases of injury, together with the addresses of ten medical men, including two professors of anatomy, who had suffered in their own families from vaccination. Concerning these remarkable evidences a leading physician, Dr. Maclean, observes:—

"Although numerous, they are nothing to what might be produced. . . . It will be thought incumbent on the vaccinators to come forward and dispute the numerous facts decisive against vaccination here stated on unimpeachable authorities, or make the *amende honorable* by a manly recantation. But experience forbids us to expect any such fair and magnanimous proceeding, and we may be assured that under no circumstances will they abandon so lucrative a practice until the practice abandons them."

We commend these prophetic words, uttered years ago, to those who look for the impartial treatment of this question at the hands of professional propagandists at the present day.

VACCINATION FAILURES IN 1817.

"However painful the duty, we feel ourselves called upon to notice the numerous and accumulating failures of cow-pox in preventing small-pox, whether in the natural way or by inoculation. Our communications on the subject have been numerous, and some of the cases do not appear to have been modified by the previous disease. It is not easy to account for these distressing occurrences, but were we to hazard a conjecture, we would venture to suggest that it is possible the virus may have become so modified by being confined altogether to the human subject that its powers of producing the necessary affection of the constitution, which only can be regarded as the test of security, may be so nearly worn out as to be no longer a certain preventive. Hence the necessity of frequently renewing the efficacy of vaccination by procuring the virus directly from its original source.

"Variola continues and spreads a devastating contagion. However painful, yet it is a duty we owe to the public and the profession to apprise them that the number of all ranks suffering under small-pox, who have previously undergone vaccination by the most skilful practitioners, is at present alarmingly great."—*London Medical Repository, pp. 57 and 95. Edited by George Burrows, M.D., F.L.S., etc., etc., and Anthony Todd Thompson, F.L.S., M.R.C.S, etc.*

ARM-TO-ARM *versus* CALF-LYMPH VACCINATION.

As the relative merits and demerits of animal and arm-to-arm vaccination are being discussed in the Indian press, I would here call attention to a few facts which appear to have escaped the attention of those who are urging the adoption of new compulsory laws in various districts. It is generally admitted that there is a growing discontent with vaccination in all quarters. In England household censuses have been made in nearly one hundred towns and districts, with the result that eighty-seven per cent. of the signatories are opposed to compulsory vaccination and sixty-eight per cent. certify that they do not believe in vaccination at all. In about ninety towns and poor-law unions the acts are a dead letter; and, owing to the sinister results of the practice, in many of our Colonies the authorities have ceased prosecuting vaccine recusants. In India, Sir John Gorst (*Times*, 17th July, 1891) informed Parliament that compulsory vaccination exists in four districts in Bengal and in 183 municipal cities and towns in different provinces. But the results cannot be considered encouraging even by the most ardent vaccine optimist. The Blue-Books of "Sanitary Measures in India" state the total vaccinations during the years 1886-89 inclusive as follows :—

1886-87,	5,265,024
1887-88,	5,552,710
1888-89,	6,099,733
1889-90,	6,161,407

The statistical abstract relating to British India gives the small-pox mortality during the same period, viz. :—

1886,	51,112
1887,	65,757
1888,	138,509
1889,	125,453

This increase of small-pox, co-incident with the rapid extension of vaccination, shows that it is a disease governed by causes entirely outside and independent of vaccination. And this opinion is confirmed by the highest authorities. Thus, in a memorandum of the "Army Sanitary Commission," published in the *Bombay Government Gazette*, Dec. 17*th*, 1885, the Commissioners say :— "The first disease in the list—namely, small-pox, which yielded an increase of 1369 deaths, or nearly sixteen-fold that of the previous year's death-rate—had assumed an epidemic state in nearly all the districts of the city; yet Bombay has an effective vaccination service, with the use of calf-lymph." In vol. xviii. of "Sanitary Measures in India," page 203, in reference to the small-pox epidemic of 1884, it is stated :—"We are thus brought face to face with the fact that, notwithstanding the existence of an active vaccination service, small-pox swept over the provinces just as if there had been none." In the same volume, referring to Madras, the Commissioners say :—"No less than seventy-four per cent. of the small-pox deaths in Madras town occurred among children under three years of age." In Punjab, "the Compulsory Act was in force in the Amritsar municipality, but here the deaths from small-pox were far more numerous than in any other town of the province." In vol. xix., page 113, is the following candid admission:—"Ten years' statistics afford no evidence that vaccination affects the usual epidemic course of the disease, and hence this fact, in the face of the extensive vaccination work of the present and past years, appears to lead to the conclusion that in its epidemic form small-pox must be met by improving the sanitary condition of the people."

On page 10 of the Report on the Annual Returns of the Civil Hospitals and Dispensaries in Madras for 1888, under the head of Canara, South, I find that while vaccination is making satisfactory progress, the number of vaccinations having increased by 8053 cases, yet "small-pox was more prevalent than usual in the district, and was epidemic in the town of Mangalore." On pp. 9 and 10, under the head of Arcot, North, it is stated that small-pox was prevalent, and that vaccination "was performed in a careless and perfunctory manner." This careless performance of the Jennerian rite is the rule in our Crown Colonies where leprosy is prevalent, as I have found by personal investigation.

According to Sir Edwin Chadwick, Dr. B. W. Richardson, and all other sanitarians of repute, small-pox is a disease due to insanitary conditions, impure water, bad drainage, dirty living, and particularly to overcrowding; and, instead of removing these conditions, the Governments of India during the past thirty years have been spending their energies, and large sums of money, in extending vaccination. Now that the arm-to-arm system has been thoroughly discredited and shown to be futile as a preventive of small-pox and fertile as a disseminator of eczema, syphilis, and leprosy, the cry of the official vaccinator is not the sensible one of "do away with vaccination," but, let us change front and resort to the calf, sheep, buffalo, donkey, or to lanoline lymph—or anything, rather than confess that the Jennerian system is a humiliating failure. It is well known that animal lymph has been a fruitful cause of the spread of disease in Europe. On June 17, 1885, an official re-vaccination with "re-generated" lymph at the Island of Rügen, North Germany, caused an infection of a loathsome eruptive skin disease *(Impetigo Contagiosa)* of 320 children and adults. The details of this sinister affair, from Dr. Koehler, of the Imperial Medical Department, are in my possession, and have been brought before the Royal Commission on Vaccination now sitting. This disaster was due to the use of virus obtained from the Government calf-lymph establishment, Stettin. In December, 1891, when in Launceston, Tasmania, I learnt that from 200 to 300 children and adults had been afflicted with ulcerative swellings and acute *septicæmia*, caused through animal vaccination in 1887, and that the law in that colony had been suspended. Animal vaccination has no claim to public attention by reason either of its safety or of its novelty. Mr. Farn, the Government Inspector of vaccine lymph, has declared before the Royal Vaccination Commission that he cannot tell by microscopic examination whether lymph is pure or not; and Dr. Robert Cory says that the admixture of lymph with blood, which occurs in the majority of cases, does not prevent its being used.

Perhaps the most remarkable official pronouncement on this subject ever made in England is that of the late President of the Local Government Board, during the debate in Parliament on Supply, July 22nd, 1887, at which the writer was present. The object of the declaration was to allay public anxiety as to the safety of the lymph supplied by the Government. Mr. Ritchie, in the course of his speech, said:—"The honourable member for East Donegal (Mr. Arthur O'Connor) said something about lymph. He said, I think, that it was the virus of modified small-pox. I cannot agree with the honourable member in his definition as to that point. I am informed that no lymph which is used for vaccination of any kind has ever, within the memory of man, passed through the human body. Dr. Jenner's first lymph was derived from an animal source; and the lymph which is now sent out is calf-lymph. None of the lymph, I say—at all events in historic times—has passed through the human body; therefore I cannot think that the honourable gentleman is in any way justified in calling the lymph modified small-pox."

Mr. Arthur O'Connor: —"What is it, then?"

Mr. Ritchie:—"I am afraid I am not qualified to give the honourable gentleman a medical opinion of what lymph is. I have told him whence it is derived, and he will see there is no ground for calling it modified small-pox."—*Hansard's Debates, 3rd Series, vol. 317, p. 1803. July 22nd, 1887.*

The chief of the Public Health Department was clearly not aware that until a comparatively recent period arm-to-arm vaccination was practically the *only method in vogue;* and at the time Mr. Ritchie's declaration was made, to the effect that none of the lymph in use had passed through the human body, at least three-fourths of the lymph in use in the United Kingdom was the variety known as arm-to-arm vaccination virus.'

DR. R. H. BAKEWELL ON THE RISKS OF VACCINATION.

In a paper read before the Auckland (New Zealand) Institute, 20th July, 1891, and printed in vol. xxix. of the "Transactions of the New Zealand Institute," Dr. R. Hall Bakewell, formerly Vaccinator-General and Medical Officer of Health for the Colony of Trinidad ; author of the "Pathology and Treatment of Smallpox ;" Fellow of the Royal Medical and Chirurgical Society of London, etc., says:—"The permanent change in the blood is quite another matter. I commenced, but have never completed, some microscopical investigations into the conditions of the infant's blood before, during, and after vaccination. It is evident that a fertile field for inquiry is open here ; and without a series of well-conducted examinations, extending over children of different races, and in different climates, no positive conclusions could be arrived at. But of one thing we are quite certain, as it does not need the aid of a microscope ; there is a large destruction of the red corpuscles during the febrile stage of vaccinia, followed by an anæmic condition. How long this anæmic condition lasts we have no trustworthy observations to tell us ; and how far it extends —that is, what is the actual loss of red corpuscles—is, as far as I know, in the same state of uncertainty. Of course, we often find parents complaining that children who were perfectly healthy before vaccination have lost colour, strength, and flesh after it, and have never recovered their previous good health. But these complaints, tinctured as they evidently are by a strong prejudice against compulsory vaccination, must be received with caution. Still, there is such a mass of evidence of this kind that it ought to be allowed some weight.

So much for the inevitable results of vaccination. The accidents of vaccination may be roughly classified under the following heads :—

1. Inflammatory : including erysipelas and other septicæmic diseases ; glandular swellings ; phagedæna, sloughing, or mortification at the points vaccinated.

2. Eruptive diseases, mostly of a pustular character, occurring with or immediately after the vaccine eruption ; eczema, herpetic eruptions, ecthyma, and impetigo.

3. The inoculation of constitutional diseases—syphilis, leprosy, tubercle.

Now, as regards the inflammatory diseases, there are some vaccinators of large experience who assert that they have never seen any ill-results of this kind arising from vaccination. Well, some people are very lucky, but they have no right to argue from their limited experience that such accidents never occur. I have been very fortunate in my midwifery cases ; I have never lost a case in my own practice for thirty-five years ; but for all that I do not deny that women die in childbirth. I have seen erysipelas more than once or twice, or a dozen times. In the West Indies it used to be common. The inflammation that followed the vaccination of coloured children was very intense, and the number of insects attacking the unfortunate children no doubt contributed to carry the germs of erysipelas to them. Glandular swellings, particularly in scrofulous children, are not rare. I had myself a case in which each vaccine vesicle was followed by mortification of the skin beneath it, and a phagedænic ulceration which required very vigorous measures to stop it. This was in a young woman during the epidemic period in Trinidad. I am not sure whether it was a primary vaccination or a re-vaccination. The latter, as is well known, causes very severe inflammation, pain, glandular irritation, and erysipelas in the majority of adults, besides severe and most oppressive febrile disturbance ; at least, this is the case at the time of epidemics, when re-vaccination is most practised.

Post-vaccinal eruptions are so very common amongst the children of the poorer classes in England that they form one of the stock arguments against vaccination.

The inoculation of constitutional diseases used to be laughed to scorn in my younger days. It was said in my hearing by Sir John Simon, K.C.B., then Mr. Simon, the Medical Officer of the Privy Council, that no such inoculation could take place without gross carelessness or unskilfulness on the part of the vaccinator. I used to be of the same opinion; but a case I saw some sixteen or seventeen years ago convinced me that an infant might look perfectly healthy, and yet be the subject of unmistakable hereditary syphilis. The evidence that syphilis has been communicated by vaccination is simply overwhelming. I may refer to the report of the Committee of the House of Commons on compulsory vaccination; the third report of the Royal Commission on vaccination now sitting in London; the work of Mr. Jonathan Hutchinson, F.R.S., late President of the Royal College of Surgeons of England, on syphilis, in which he devotes a chapter to the description of vaccinal syphilis; and my own experience in this colony and elsewhere. I have seen three cases in this colony alone.

"On my return to Trinidad I had to encounter an epidemic of small-pox which spurred us on to vaccinate right and left, and to revaccinate all who would submit to the operation. But so firmly fixed was the belief of the people that vaccination from a child of a leprous family would be a possible cause of the vaccinated persons becoming leprous, that not even the fear of such a terrible epidemic of small-pox as was then going on would induce them to allow themselves or their children to be vaccinated from any vaccinifer in whose family any member was a leper. And then, to my astonishment and dismay, I found that there was hardly a Creole family in the island—white, coloured, or black—free from the taint of leprosy."

LEPROSY IN MADEIRA.

Dr. Julius Goldschmidt, Medical Superintendent, Lazaretto Hospital, Madeira, who has made the pathology and treatment of leprosy a special subject of study, sent a communication to the Leprosy Investigation Committee (*Journal*, Dec., 1891), in which he says:—"As to treatment, I have tried the most varied applications without any lasting benefit." In his pamphlet, "Die

Lepra auf Madeira," illustrated with reprints of photographs of hospital, and other Madeira cases (of which he kindly gave me a copy), Dr. Goldschmidt refers to his inoculative experiments with tuberculin and other drugs, and says :—"As far back as eleven years, I tried to inoculate the anæsthetic form on the tubercular one, without success." I ventured to point out to Dr. Goldschmidt how medical testimony showed that, while syphilis and leprosy were difficult to inoculate direct from the diseased to those free from these diseases, the evidence that these and other diseases were readily inoculable by means of an intermediary host such as vaccine virus was now overwhelming. Dr. Goldschmidt went on to say that he had seen such terrible results following arm-to-arm vaccination, that he now resolutely refused to use this vaccine, no matter how healthy in appearance the vacciniter might be ; and he restricted his practice entirely to imported animal lymph. The Lazaretto at Funchal, which dates from the sixteenth century, is a curious structure, and contains a series of cubicles or boxes, one of which was formerly allotted to each patient, who was strictly isolated under the then prevailing belief that the disease was spread by simple contact. I was surprised to notice that a poor little boy about six years old, free from the disease, was living amongst these afflicted leper patients because he had no other home. There is always danger of accidental inoculation, particularly with children ; and it is to be hoped that some benevolently disposed resident of Funchal (and I know there are many such) will take steps to rescue him from such unhappy surroundings. It may be imagined that the inmates formerly passed a miserable existence in their dark and prison-like cells. Dr. Goldschmidt states that there are ten times as many lepers at large in Madeira as are brought to the asylum, and, having regard to the number of lepers I saw in Funchal and neighbourhood, this is by no means an exaggerated estimate. Vaccination is compulsory in the Island of Madeira, but is enforced only when smallpox prevails, and is then performed in a careless, perfunctory, and hurried manner, with the result of spreading various diseases amongst an ill-conditioned, poorly-fed population, who for the most part live under unfavourable sanitary surroundings. Dr. Goldschmidt is of opinion that leprosy in Madeira is largely due to unwholesome food and impure water, there being no proper means of providing pure water to the poorer part of the population ; and while he considers that leprosy itself is not hereditary, he thinks he has established the fact that certain conditions which predispose the individual to the disease are hereditary. He is by no means satisfied with the work of the Leprosy Commission, and advocates the organisation of an International Leprosy Congress in London, where the hotly-disputed etiological theories could be exhaustively threshed out. Dr. Goldschmidt estimates the number of lepers in Madeira at seventy, or six to 10,000 inhabitants, and is of opinion that it is not increasing.

SIR JAMES PAGET ON SURGICAL PATHOLOGY.

"After the vaccine and other infectious or inoculable diseases, it is, most probably, not the tissues alone, but the blood as much as or more than they, in which the altered state is maintained ; and in many cases it would seem that, whatever materials are added to the blood, the stamp once impressed by one of these specific diseases is retained ; the blood, by its own formative power, exactly assimilating to itself, its altered self, the materials derived from the food.

"And this, surely, must be the explanation of many of the most inveterate diseases ; that they persist because of the assimilative formation of the blood. Syphilis, lepra, eczema, gout, and many more, seem thus to be perpetuated : in some form or other and in varying quantity, whether it manifests itself externally or not, the material they depend on is still in the blood ; because the blood constantly makes it afresh out of the materials that are added to it, let those materials be almost what they may. The tissues affected may (and often do) in these cases recover ; they may have gained their right or perfect composition ; but the blood, by assimilation, still retains its taint, though it may have in it not one of the particles on which the taint first passed ; and hence, after many years of seeming health, the disease may break out again from the blood, and affect a part which was never before diseased."—*Lectures on Surgical Pathology, 4th ed., p. 39. 1870.*

DR. M. D. MAKUNA'S MEDICAL VACCINATION CENSUS.

Prior to the appointment of the Royal Commission on Vaccination in 1889, the latest professional examination of the vaccination question was in the form of a medical inquiry, initiated by Dr. Montague Makuna, late superintendent of the Fulham Small-pox Hospital. Its object was to remove, if possible, the widespread feeling of mistrust prevalent, in various parts of the United Kingdom, as to the benefits and safety of vaccination, which had been discredited in public estimation by reason of the failures and mischiefs of the operation, with details of which the press was flooded. The inquiry was made by a committee of thirty medical gentlemen, most of whom were vaccine specialists. The meetings were held in the Council Chamber of Exeter Hall, London, under the presidency of Dr. C. R. Drysdale, senior physician of the Metropolitan Free Hospital, a gentleman who has devoted much time to this important question. Dr. Drysdale promised the Committee to read a paper on the results of the inquiry. The proceedings were reported in the *Midland Medical Miscellany* and in the *Medical Press and Circular*. A circular letter, with copy of the report of the first meeting of the Committee, was sent to members of both Houses of Parliament.

At the first meeting of the Committee, held on 15th February, 1883, Dr. Makuna, in explaining the objects of the proposed inquiry, referred to the opposition to vaccination; an opposition which has become intensified since vaccination has been made compulsory. A medical inquiry was, therefore, considered indispensable, and it was anticipated that the evidence disclosed in favour of vaccination would be so unanimous and conclusive, as to effectually restore public confidence in the practice, and to put an end to all opposition. The chairman, Dr. Drysdale, said he considered the proposed inquiry would be of great value to the profession and to the public, and expressed a desire that the Local Government Board and other authorities should be requested to co-operate.

The attention of the Local Government Board, the British Medical Association, the Epidemiological Society, the Medical

Officers of Health, the Royal College of Physicians, as well as ambassadors, consuls, etc., was specially called to the inquiry, and they were requested to contribute facts and information from all parts of the world.

A circular was drawn up and approved by the Council, and sent to 4000 medical practitioners, a considerable portion of whom were public vaccinators, medical officers of health, and vaccine specialists. The circular elicited 384 answers, and the results were published in a pamphlet entitled "Transactions of the Vaccination Inquiry."

An analysis of the answers (made by Mr. Thomas Baker, Barrister-at-Law) shows that the seven questions submitted have been answered by 384 medical men, of whom 102 are public vaccinators, vaccine specialists, medical officers of health, or officials.

The following is the third, and one of the most important, submitted to this medical inquisition, viz. :—

What diseases have you in your experience known to be conveyed or occasioned or intensified by vaccination? To this question 13 give no answer, and 139 answer "None,"[1] but many qualify this reply by the words, "in my own practice," "direct," "not serious," "personal," etc. On the other hand, the list of mischiefs (many fatal) includes the following, as recorded by 232 medical witnesses. (This enumeration has been checked by the Rev. Isaac Doxsey, F.S.S., and Mr. J. H. Lynn :—)

	Witnesses.		Witnesses.
Abdominal phthisis,	1	Boils,	8
Abscesses,	11	Bronchitis,	1
Angeioleucitis,	2	Bullæ,	1
Arm disease needing amputation,	1	Cancer,	1
Axillary Bubo,	1	Cellulitis,	5
,, gland, enlargement of,	1	Convulsions,	4
Blindness,	1	Diarrhœa,	4
Blood poisoning (fatal),	1	"Died,"	1

---

[1] No. 17 says: "Two deaths from erysipelas occurred after vaccination in my practice, both commencing on the ninth day." Therefore, that so many answer, "None," may be accounted for by the fact that public vaccinators commonly do not see the vaccinated child after the eighth day.

Witnesses.		Witnesses.	
Diseased Bones,	1	Phagedænic action,	1
Diseased Joints,	1	Phlegmon,	2
Dyscrasia,	1	Pityriasis,	1
Ecthyma,	1	Pneumonia,	1
Eczema,	60	Prurigo,	3
Eruptions,	5	Psoriasis,	1
Erysipelas,	120	Pyæmia,	7
Erythema,	22	Pyrexia,	1
Gangrenosa,	3	Rickets,	1
General Debility,	1	Scald head,	1
Herpes,	3	Scarlatina,	3
Impetigo,	7	Scrofula,	9
Inflammation,	10	Septicæmia,	1
Latent diseases developed,	2	Skin disease,	21
Lichen,	2	Struma intensified,	4
Marasmus,	1	Syphilis,	43
Meningitis,	2	Tuberculosis,	1
Mesenteric disease,	1	Ulceration,	6
Œdema,	2	Varioloid,	1
Paralysis,	1		

The following *qualified* admissions are also made, viz.:—

Eczema,	4	Nettle rash,	1
Erysipelas,	6	Syphilis,	10

On the 19th of April, 1886, the results of this medical census were laid before the Right Hon. James Stansfeld, then President of the Local Government Board, and in May a copy of the analysis was sent to him at his request. And on the 25th June, and on the 2nd July, 1890, I called the attention of the Royal Commission on Vaccination to this important inquiry, and presented an analysis of the results for its consideration. The facts ought, in the interest of the public safety, to be widely disseminated through the press, and made known to magistrates, boards of guardians, and others concerned with the enforcement of the Vaccination Acts.

It has, I think, been clearly proved on the evidence of medical specialists, that to the long catalogue of diseases conveyed at the point of the lancet in vaccination must now be added that of leprosy. This terrible indictment has been denied again and again by those who have not taken the trouble to investigate the facts for themselves, and have shut their eyes to the facts revealed by others. Is it reasonable to suppose that negative statements, however confidently made, can destroy the positive testimony of careful investigators such as Dr. John D. Hillis, Professor W. T. Gairdner, Dr. Edward Arning, and Dr. S. P. Impey? The fact that syphilis is communicable by vaccination was emphatically repudiated by a President of the Local Government Board not long ago, and the terrible Algiers vaccine disaster in 1880 was officially proclaimed an impossible occurrence. The fact that leprosy may be communicated by vaccination is now reluctantly admitted even by those who are most anxious to clear the rite from this reproach. In concluding a defence of vaccination from the charge referred to in No. 4 of the *Journal of the Leprosy Investigation Committee*, Dr. Beaven Rake says : " It is evident that the risk of transmission of leprosy by vaccination is so small that, for all practical purposes, it may be disregarded." Dr. P. Abraham says : " The possibility of an occasional accidental inoculation of the disease by vaccination might be admitted ;" and Dr. C. F. Castor, in a paper defending vaccination from this stigma, observes : " The opinion expressed that vaccination from a tainted source will produce the disease (leprosy) is, I believe, a true one."

IS VACCINATION A PREVENTIVE OF SMALL-POX?

When presenting his evidence before the Royal Commission on Vaccination, July 2, 1890, the author produced the following letter which he had received from the late Herr G. F. Kolb, of Munich, Member Extraordinary of the Royal Statistical Commission of Bavaria, and one of the distinguished statisticians of Europe. It is dated January 22, 1882. Dr. Kolb says:—"From childhood I have been trained to look upon the cow-pox as an absolute and unqualified protective. I have, from my earliest remembrance, believed in it more strongly than in any clerical tenet or ecclesiastical dogma. The numerous and acknowledged failures did not shake my faith. I attributed them either to the carelessness of the operator, or the badness of the lymph. In the course of time the question of vaccine compulsion came before the Reichstag, when a medical friend supplied me with a mass of pro-vaccination statistics, in his opinion conclusive and unanswerable. On inspection I found the figures were delusive, and a closer examination left no shadow of doubt in my mind that the so-called statistical array of proof was a complete failure. My investigations were continued, but with a similar result. For instance, in the kingdom of Bavaria, into which the cow-pox was introduced in 1807, and where for a long time no one except the newly-born escaped vaccination, there were in the epidemic of 1871 no less than 30,742 cases of small-pox, of whom 29,429 had been vaccinated, as is shown in the documents of the State Department. When, with these stern proofs before us of the inability of vaccination to protect, we reflect upon the undeniable and fearful mischief which the operator so often inflicts upon his victims, the conclusion forces itself upon us that the State is not entitled either in justice or in reason to put in force an enactment so directly subversive of the great principle of personal right. In this matter State compulsion is, in my opinion, utterly unjustifiable."

MEDICAL DENIALS AND ADMISSIONS.

The four numbers of the *Journal* issued by the Leprosy Investigation Committee contain a large amount of testimony from authorities in every part of the globe. A conspicuous feature, and one which has been commented upon in the press, is the confusion and contradiction of medical opinion, and this confusion is not restricted to theorists who have never visited countries where leprosy prevails, but is exhibited amongst superintendents of leper hospitals, well-known dermatologists, and eminent general practitioners. Most of the theorists are able to cite facts in support of their several beliefs, and maintain their conflicting opinions with equal confidence. Some of the most prominent of these views have been dealt with, but it would be beside the scope of this work to refer to them all. Beyond the general admission of the alarming increase of the disease, there is but one point upon which there is any approach to a consensus of opinion, and that is, that leprosy is a disease communicable by inoculation. Conversing with directors of leper institutions, experienced practitioners, and careful observers in countries where leprosy prevails, whatever their particular theories about it, I have found no one who denies that it is inoculable. In some instances, when pressing the logical issue of these answers I have said, "Then the disease is also communicable by vaccination," this has often been admitted, though sometimes with hesitation and reserve, and generally accompanied with the proviso that leprosy cannot be invaccinated if pure lymph only is employed and the operation be skilfully performed. Even if this were true, how pure lymph can be selected in tropical countries where the disease is generally of slow incubation, and does not manifest itself for years, and where the vaccinifer is never properly examined, is a mystery not explained. Moreover, while the State can enforce vaccination, it cannot compel the use of healthy vaccinifers, nor enforce careful operation. In nearly all leprous countries arm-to-arm vaccination still furnishes the chief sources of supply. The perennial cry of public vaccinators is that the lymph is "unsatisfactory." Animal lymph is often attended with excessive inflammation, and the practitioner is obliged to dilute it with glycerine, lanoline, and other substances, and its use is much more expensive. Moreover, a good deal of the so-called animal lymph in vogue is really only arm-to-arm vaccine, inoculated into calves, buffaloes, sheep, and donkeys, and partakes of the diseases both of man and of animals. Of the many cases of ulcerative and of fatal vaccination which have come under my notice during the past twenty years not a few have been due to the use of carefully-selected animal vaccine.

Dr. Robert Cory, Medical Director of the Government Calf Lymph Establishment, London, testified before the Royal Commission, November 17, 1889 (Q. 4390), that out of 32,000 cases there were 260 returned with sore arms, and 38 with eruptions. Then there were 16 cases of erysipelas, and nine of axillary abscesses, and (Q. 4392) eight deaths were reported to the Station of children who had been vaccinated with animal lymph. The same witness testified (Q. 4369) that lymph taken from the cow leads to greater inflammation, and has a greater tendency to produce ulceration, than lymph which has been humanised.

HOW LEICESTER DEALS WITH VARIOLOUS OUTBREAKS.

The success of what is known as the "Leicester Experiment" has created considerable public interest both at home and abroad, and, in response to repeated inquiries, I am able to furnish the following particulars from the pen of Mr. Councillor Biggs, of Leicester. Mr. Biggs was for several years a member of the Board of Guardians, and is now a member of both the Sanitary and Small-Pox Hospital Committees. He has had much to do with the substitution of sanitary amelioration and isolation for the now discarded system (so far as Leicester is concerned) of vaccination.

Our procedure in the notification of a case of small-pox may (he says) be described in the words of our Chief Sanitary Inspector, Mr. F. Braley: —

"When a case is reported, I at once go to the infected house, and try to ascertain where the disease was contracted, where the patient has been working, where he has been visiting, and his movements generally for the last ten or twelve days. I also make a point of seeing all persons who have visited the infected house during the time stated; in addition, I visit all factories and workshops where other members of the family have been employed; and by this means have been able to get cases removed when the first symptoms of the disease appeared.

"Immediately on the removal of the patient, I superintend the fumigation of the house with sulphur; liquid disinfectants are used freely in the drains and about the yard, and the ashpit is emptied and disinfected; the next day the bedding is taken to the disinfecting chamber and subjected to the hot-air process.

"Up to the present time I have succeeded in getting almost every person connected with the infected houses into quarantine. In a very few cases I have experienced opposition.

"The above represents practically all we do of a special character beyond the ordinary treatment in cases of supposed infection. Those who are prevailed upon to go into quarantine usually remain for fourteen days, the period within which small-pox is supposed to incubate after infection. So far from the authorities having to resort to harsh measures to enforce quarantine, this period of rest is made to be of so pleasant and agreeable a character that, at its expiration, many have been reluctant to leave the hospital.

VACCINATION OPTIONAL.

"This fact disposes of an accusation which is constantly being hurled at Leicester, namely, that we not only forcibly seize those who have been in contact with small-pox cases, and compulsorily detain them in quarantine, but that, whilst they are there, we compel them to submit to vaccination or re-vaccination.

"Let us examine as to how far this favourite theory of our opponents has any foundation in fact.

"Of fifty-five persons who voluntarily went into quarantine during the three years 1886-88, only twelve were vaccinated or re-vaccinated whilst in quarantine. If to these twelve we add three others who underwent the operation immediately before entering the hospital, there remain forty persons, or, 72.7 per cent. of the above fifty-five, who were neither vaccinated nor re-vaccinated during the quarantine period.

"Applying the same percentage to the 128 persons quarantined before 1886, we find a further number of thirty-five persons who voluntarily submitted to the operation, who, when added to the other fifteen, make only fifty for the extended period. Thus, out of a total number of 183 persons who have passed through the quarantine wards since the introduction of this system in 1877, no fewer than 133 were neither vaccinated nor re-vaccinated whilst in the hospital.

"During this period from fifteen to twenty persons absolutely refused to go into quarantine at all, and we had no power to compel them.

"Nearly all those who were quarantined belonged to the poorest classes, and to these a fortnight's holiday with free board and good, if not comparatively luxurious, living, would prove to be no mean attraction. Those who had good homes remained there in preference to going to the hospital. Thus no infringement whatever of personal liberty has taken place against those who have put themselves for a time under the 'Leicester Method' of treatment; unless, indeed, the gastronomic allurements above referred to might have proved an inducement to some to voluntarily yield up their personal liberty for a time.

SUCCESS OF THE LEICESTER METHOD.

"After the subsidence of the great small-pox epidemic of 1871-73, which caused 360 small-pox deaths, when the town was thoroughly well vaccinated, up to the year 1889, which was the last year for which I could prepare statistics for the Royal Commission on Vaccination —that is, during the sixteen years from 1874 to 1889, inclusive—no fewer than thirty-three importations—mostly from well-vaccinated districts—and a large number of successive outbreaks of small-pox, were successfully stamped out. The town was thus saved from the further spread of the disease, with its possible ravages, by the 'Leicester Method' of treatment, *without recourse to vaccination*, and also without the slightest approach to arbitrariness on the part of the authorities, or any infringement of personal liberty.

ECONOMY OF THE "LEICESTER METHOD."

"It is sometimes assumed that this 'Leicester Method' of isolation, quarantine, disinfection, and sanitation is so expensive as to be practically prohibitive.

"On the contrary, our 'Leicester Method' is extremely economical as well as effective. Besides, it is now well-known that, however thoroughly a community is vaccinated, so little reliance is placed upon this supposed safeguard, that on the outbreak of small-pox recourse is at once had to the very measures which have been so persistently decried when used to the salvation of unvaccinated Leicester.

"From 1874 to 1889 the cost of public and private vaccination at Leicester was not far short of £10,000 (being about £9818 2s. 11d.). During the same period the cost of quarantine, including compensation for destruction of infected clothes, bedding, disinfectants, etc., was under the modest sum of £500 (or about £488 11s. 2d.). This represents a saving in favour of our Leicester method, as against vaccination, of over £9000 in the course of sixteen years. This £9000 was completely thrown away, to say nothing of the impaired vitality and spread of disease which vaccination necessarily implies. The £500 cost of quarantine, etc., did all the effectual work of saving the town from the ravages of small-pox threatened by the thirty-three importations, and absolutely averted the real danger implied by the occurrence of 116 small-pox cases in the midst of our crowded population.

JUSTIFIED BY RESULTS.

"Perhaps it will not now be out of place to briefly enumerate the substantial reasons which justify the Leicester people in the course they have pursued in respect to vaccination, and in adopting sanitation as their defence in the conflict with zymotic disease.

"Taking the groups of years dealt with by me before the Royal Commission on Vaccination, our average annual small-pox death-rate during 1853-57, with a moderate amount of vaccination, was only 91 per million population. But when vaccination had been continually and largely practised for a quarter of a century, and had reached over 90 per cent. to the annual births, and when, of course, its assumed protective power should have been greatest, our small-pox death-rate had progressively risen to an annual average of 773 per million population in 1868-72. Since that time vaccination has rapidly declined in the Borough, now being only about 2 per cent. of the births, and small-pox mortality has disappeared from our midst.

SAVING OF LIFE IN LEICESTER.

"Our death-rate from the seven principal zymotic diseases, namely, small-pox, measles, scarlet fever, diphtheria, whooping-cough, common fevers (typhus, typhoid, and continued fever), and diarrhœa, averaged annually for the five years 1868-72 no fewer than 6852 per million living, with over 90 per cent. of primary vaccinations to births. This is the highest vaccination rate and zymotic death-rate we have ever had recorded for Leicester. In 1888-89, when primary vaccinations were only about 5 per cent. of the births, the zymotic death-rate had fallen to only 2304 per million. On our Leicester population alone this would mean a saving of nearly 680 lives each year.

"Without going into unnecessary details, I may observe that

the improvement in our general death-rate amongst children shows equally remarkable results. With over 90 per cent. of primary vaccinations to births in 1868-72, our death-rate from all causes, of children under five years of age, was 107; under ten years, was 61; and under fifteen years was 45 per 1000 living under each of those ages respectively. While in 1888-89, with only about 5 per cent. of primary vaccinations to births, each of these death-rates had fallen enormously. The death-rate under five years had declined to 63, that under ten years to 35, and that under fifteen years to 25 per 1000 living at each of the given ages respectively.

"This would represent a saving of about 880 lives under five years, of about 988 lives under ten years, and of about 1080 lives under fifteen years of age, inclusive and respectively for each year in Leicester.

"When it is remembered that the claim put forward for vaccination is its preservation of the younger lives, especially those under five years of age, the life-saving result of the 'Leicester method,' as shown above, is particularly striking. And it proves, unmistakably, that our watchword, 'Sanitation,' carries with it far more potency to deal with zymotic disease and with small-pox than the now discredited cry of 'Vaccination.'

THE GENERAL HEALTH OF LEICESTER.

"Once more our general death-rate, that is, our death-rate from *all causes* and at *all ages*, gives results no less important. In 1868-72, when vaccination had reached its climax in Leicester, our death-rate was about 27 per 1000 of the living population, being nearly 5 per 1000 above the general death-rate for all England and Wales. In 1888-89, when vaccination had virtually ceased to be practised in the town, notwithstanding our disadvantageous geological and geographical position, in a valley, with one of the most sluggish rivers in England and a clayey and impervious water-logged subsoil, our incomplete, and therefore inadequate, drainage, our death-rate had fallen so rapidly, with declining vaccination, that it had actually fallen below the general death-rate of England and Wales. The death-rate for England and Wales was 17·9 for 1888-89, and that for Leicester 17·5, or 5·1 gain in favour of Leicester in less than twenty years.

"These figures as compared with times of high vaccination mean an additional saving of about 1400 lives each year in Leicester alone, above the normal rate of saving in England and Wales. If this extra gain could be similarly achieved by the cessation of vaccination in the population of the whole country, other things being equal, it would mean an enormous saving of life beyond that which has actually been effected. The population of the United Kingdom for 1888-89 was estimated by the Registrar-General to be over 37,000,000. On this population an annual saving of about 189,000 lives would be effected. Even allowing an ample margin for possible errors in the calculations of the Registrar-General, these figures are sufficiently momentous to claim serious consideration.

"When it is borne in mind that England and Wales include all the rural districts, where the death-rate is very low, and that here our people are chiefly an artizan and manufacturing population whose circumstances are ever inimical to the health of the younger lives, Leicester's progress from being one of the unhealthiest of towns to its present proud position must be acknowledged to be marvellous.

"With such remarkable results before us, the Leicester people can calmly await the verdict of thoughtful minds, assured that their course of action in rejecting vaccination, and their reliance upon sanitation, will in the long run break down existing prejudice, and that it will ultimately receive general approval and adoption."

It need only be added that a full report of the Leicester system has been presented by Mr. Biggs before the Royal Commission on Vaccination, which will be found in the fourth report of the evidence.

COW-POX AND VACCINAL SYPHILIS.

The following extracts from Dr. Creighton's "Natural History of Cow-pox and Vaccinal Syphilis" (London, 1887) express the conclusions of his historical inquiry—namely, that cow-pox is a disease resembling small-pox only in name, but resembling the great pox both in name and in reality, and that so-called vaccinal syphilis is only cow-pox in its original form, as the milkers used to experience it :

"The real affinity of cow-pox is not to the small-pox but to the great pox. The vaccinal roseola is not only very like the syphilitic roseola, but it means the same sort of thing. The vaccinal ulcer of every-day practice is, to all intents and purposes, a chancre. It is apt to be an indurated sore when excavated under the scab; when the scab does not adhere, it often shows an unmistakable tendency to phagedæna. There are doubtless many cases of it where constitutional symptoms are either in abeyance or too slight to attract notice. But in other instances, to judge from the groups of cases to which inquiry has been mostly directed, the degeneration of the vesicle to an indurated or phagedænic sore (all in its day's work) has been followed by roseola, or by scaly and even pemphigoid eruptions, by iritis, by raised patches or sores on the tonsils and other parts of the mouth or throat, and by condylomata (mucous tubercles) elsewhere." (Page 155.)

"The first duty of everyone is once for all to disabuse his mind of Jenner's invention of the name *variolæ vaccinæ* for cow-pox. The affection of the cow's udder was long recognised by common folk as a pox in the original and classical English sense of the word; the name of it in Norfolk was pap pox. No one had dreamt of discovering any resemblance in it to the pustules of the foreign contagious skin-disease which came to be called the small-pox, until Jenner, by a masterstroke of boldness and cunning, placed the Latin name *variolæ vaccinæ* first on his title page, as if he were merely expressing in scientific form the universally accepted meaning of the colloquial name. There was no candid or overt attempt, in the body of his essay, to justify that daring innovation; most of his readers from that time to this have hardly realised that it was an innovation at all, for the reason that Jenner adroitly left his title page to justify itself. His trumped-up name somehow passed without challenge, except for a grammatical objection on the part of Pearson, and a general criticism by Moseley; and although the want of likeness, still more in circumstances than in form, between the pustules of small-pox and even the modified kind of inoculated cow-pox vesicle, has been pointed out in elaborate detail by several writers, and ought, indeed, to be so obvious to any one as not to need pointing out at all, yet the Jennerian fable of *variolæ vaccinæ* continues to be the creed of the medical profession." (Page 157.)

"The rational theory of the Morbihan disaster (of vaccinal syphilis in 1866) is that ulceration, followed by induration and (or) phagedæna, is part of the natural history of cow-pox infection; that it is nearly always latent, or kept in check; that in some circumstances it may be brought out or reverted to; that these circumstances, in the particular epidemic, were the date and number of the vesicles raised on the vaccinifer, and the draining of their lymph to the last drop, so as to vaccinate an enormous number;

and lastly, that a continuous reproduction of lymph from that stock tended to confirm and even to intensify the reawakened powers of the cow-pox matter, as evidenced by the more decided 'syphilitic' character of the secondaries (mucous patches on the tonsils) in two cases of the last group." (Page 140.)

" The origin of vaccinal syphilis remains, as Bohn says, 'shrouded in mystery.' Readers who have followed my argument hitherto will not be surprised if now I claim the phenomena of so-called vaccinal 'syphilis' as in no respect of venereal origin, but as due to the inherent, although mostly dormant, *natural-history characters of cow-pox itself.*" (Page 124.)

THE AUTHOR'S PERSONAL STATEMENT OF THE RESULTS OF VACCINATION.

During the past twenty-two years it has been my experience to travel in all parts of the United Kingdom, from Land's End to the Shetland Islands, and in almost every state in Europe, from the Mediterranean to the North Cape, in countries intervening between the Tagus in the west, and the Volga, Danube, and Bosphorus, in the east; also in Morocco, Algeria, Upper and Lower Egypt, Asia Minor, Upper and Lower Canada, Nova Scotia, and most of the states and territories of North America; also in Venezuela and British Guiana, South America, in the Windward and Leeward Islands, the French and Danish West Indies, in the archipelagoes of Greece and Hawaii, the island of Ceylon, in Tasmania, New Zealand, the colonies of Australia, and in South Africa.

In nearly all these countries I have made it my business to inquire into the methods and results of vaccination, procuring information from public officials and from intelligent private individuals, and I have hardly ever inquired without hearing of injuries, fatalities, and sometimes wholesale disasters, to people in every position in life, and these have occurred from the use of every variety of vaccine virus in use. My informants have included governors, chief magistrates, consuls, professors of medicine and surgery in Continental universities, members of legislative assemblies, superintendents of leper asylums, editors of medical and hygienic journals, chiefs of military and general hospitals, presidents and medical officers of state and colonial health departments, superintendents of small-pox hospitals, clergymen of all denominations, missionaries, heads of educational establishments, and the best informed amongst old residents in the places visited.

In one country it was my privilege to be furnished with a general letter of introduction from a minister of State (since Prime Minister), which gave me access to all the official and medical authorities. Often the fatality described to me has befallen the infant of a poor mother, who with dread forebodings in her mind has tried to shield her offspring from the vaccinator's lancet as long as she could, and, like a fugitive slave, only surrendered to the minister of the law when overtaken in pursuit or her place of refuge discovered ; or, like that of a distinguished Moslem (Suffey Bey Adem), my travelling companion in 1884 from Damascus to Beyrout, who had lost a daughter, a nephew, and a niece (vaccinated together about a year before our interview), all of whom died of the operation, after the most acute suffering. At other times I have seen stalwart soldiers and post-office officials seriously injured, and in more than one instance crippled and ruined for life, by compulsory re-vaccination. I have personally investigated vaccine disasters at two military hospitals, one in Europe and the other in Africa, where, in one case, three, and in another case thirty soldiers ultimately died of the operation, and more than twice this number were seriously, and, in most cases, permanently injured. In Australasia I have personally inquired into a case of wholesale disaster—of acute septicæmia, exhibited by terrible ulcerations following vaccination with calf lymph—to several hundred persons, and have seen the sad consequences in permanently ruined health. I have received several thousand written statements from parents, who allege that their children have been seriously or fatally injured by vaccination. I have proved beyond doubt, by personal inquiries in various countries where leprosy is increasing, that the increase is largely due to vaccination, and have furnished the testimonies of numerous medical authorities, and of official reports (all mention of which has been omitted from our leading medical journals), in support of these incriminating allegations. These facts have been detailed by me in the *Times*, *Nonconformist*, *Echo*, *Leeds Mercury*, *Manchester Guardian* and *Examiner and Times*, *Leicester Post*, *Newcastle Leader*, *Scottish Leader*, *Cardiff Daily News*, *Gloucester Citizen*, *Hospital Gazette*, *The Tocsin*, *Journal d' Hygiène* (Paris), *Birmingham Gazette*, *The Vaccination Inquirer*, and other influential and well-known English, American, and Colonial journals ; and some of them were quoted by me, with chapter and verse, before the Royal Commission on Vaccination, now taking evidence in London, and will be found in the third official report of the proceedings.

I may also mention that numerous facts of a sinister character were contributed by many of the delegates representing the leading European States at the International Anti-Vaccination Congresses held in Paris, Cologne, Berne, and Charleroi, the reports of which have been published and presented to the chiefs of Governments, and of Public Health Departments in all countries. Not only have the facts been submitted to Continental Ministers of State, and to successive Presidents of the Local Government Board in England, but in December, 1890, I laid them before Mr. Langridge, Chief Secretary to the Government of Victoria, Australia, and before leading officials in other Colonies. It seems to me, therefore, that, in view of these experiences, and in the presence of such unimpeachable facts, the opposition which has arisen, and is growing daily in nearly all countries, is a commendable and patriotic struggle, which should be encouraged in every possible way. The laws (often cruelly enforced), which compel the parents of this and other countries to put the health and lives of their offspring into the hands of irresponsible State officials, with the alternative of severe and not seldom ignominious punishments, is a grave national blunder, and constitutes a species of tyranny wholly indefensible ; and it behoves every good citizen to endeavour, by every constitutional means, in the interests alike of justice, of individual and parental rights, and in defence of the public health, and of our helpless children, to get these laws completely and permanently extinguished.

INDEX.

ABBOTT, DR. S. W., on leprosy and diet, 227.
Abercromby, Dr. A., 73; on contagion, 95; on inoculation, 111; on vaccine lymph, 214.
Abraham, Dr. P. S., on causation, 233; paper read by, 65; on vaccination and leprosy, 236-8; circular in *Lancet*, 252.
Accidental inoculation, 103, 111.
Ackworth, Dr. H. A., 75, 93, 108; Matunga Leper Asylum, 158; on diagnosis, 260.
Acquired leprosy, Mr. J. Hutchinson on, 235.
Africa, South, increase of leprosy in, 63-73; miscellaneous evidence concerning leprosy in, 109-12; vaccinal diseases in, 268-76; Consul Siler's report from Cape Town, 327; report from Cape of Good Hope, 341-4.
Agua de Dios, leprosy in, 89.
Allen, Dr. C. W., 22, 39; opinion of, 114.
Allison, Dr., 176.
Alston, Dr., 137.
America, leprosy in, 38, 325.
Analysis of statistics, 258.
Anæsthetic leprosy, 126, 127, 159, 171, 173, 191-9; Professor Hyde on, 325.
Anderson, Prof. M'Call, 139, 144.
Anderson, case of J., 116.
Animal lymph, objections to, 181.
Antigua, leprosy in, 31, 122, 147.
Anti-leprosy Association (Bombay), 79.
Antilles, 35; case of Sister A, 101.
Antioquia, leprosy in, 47.

Anti-Vaccination Congress, 391.
Appendix, 353.
Argus, The Cape, on the increase of leprosy, 69, 70.
Arning, Dr. E., leprosy in Hawaii, etc., 10, 41, 123, 127, 156, 158-9, 163.
Ashmead, Dr. A. S., leprosy in Japan, 129.
Asylum, leper, at—
 Abohivaraka, 45.
 Agua de Dios, 47.
 Barbados, 28.
 Bhopal, 333.
 Bombay (Matunga), 75.
 British Guiana, 174.
 Ceylon, 285.
 Contratacion, 47.
 Damascus, 37.
 Funchal (Madeira), 374.
 Gorchum, 48.
 Gotemba, 36.
 Havana, 96.
 Hendala, 78.
 Ilafy, 45
 Isle of Goats, 57.
 Jamaica (Spanish Town), 35.
 Jerusalem, 292.
 Kalihi, 44.
 La Disirade, 35.
 Maracaibo, 292.
 Mauritius, 45.
 Molokai, 40.
 Mucurapo, 29.
 New Brunswick (Tracadie), 45, 229, 326-7.
 Norway. 27.
 Pic des Morts, 57.
 Robben Island, 277.
 Sandy Point, 32.

INDEX.

Asylum, St. Kitts, 31.
  Sydney, N.S.W., 54.
  Trinidad, 30.
  Venezuela, 53, 292.
Atherstone, the Hon. Dr., on increase of leprosy, 68, 110; on heredity of leprosy, 225.
Atkinson, Dr W. B., on contagion, 87, 116; on inoculation of leprosy, 87, 116.
Auckland, leprosy in, 54.
Australia, leprosy in, 53.

BABINEAU, MR. T. A., on treatment of leprosy, 327.
Bacillus of leprosy in saliva, 95, 106; Dr. W. B. Atkinson, on, 116; in lymph, 159, 199, 205, 207.
Baker, The Rev. Canon, on leprosy, 64-65; and inoculation, 110.
Bakewell, Dr Hall, increase of leprosy in Trinidad, 28; on leprosy and vaccination, 130, 132, 135; on risks of vaccination, 371-373.
Baltic Provinces, leprosy in, 25, 221.
Barbados, leprosy in, 29.
Bayne, Mr. C. G., communication from, 77.
Bechtinger, Dr., on invaccinated leprosy, 207.
Begum of Bhopal, 333.
Besnier, Dr., report on leprosy in French Colonies, 59.
Bibliography of works referred to, 402.
Bidie, Surgeon, address by, 76.
Bisset, Sir John, on the dissemination of leprosy, 275.
Black, Dr. R. F., letter on leprosy and vaccination, 138.
Blanc, Dr. H. W., on leprous vaccination, 179.
Blaney, Dr., statement by, 74.
Bokhara, leprosy in, 26.
Bolton, Dr., of Mauritius, case of leprosy and vaccination reported by, 203.
Bombay Guardian on leprosy and vaccination, 184.
Boon, Dr., report of, 31-33.

Boral, Dr. S N., on vaccinal contagion, 182.
Bowerbank, Dr., on leprosy and vaccination, 154.
Boyacá, leprosy in, 46.
Brazil, lepers in, 49.
British Guiana, leprosy in, 49.
British Medical Journal on leprosy in Livonia, 218; Mr. Hutchinson's paper, 235.
Brodie, Surgeon-General, on syphilis and vaccination, 190.
Brown, Dr. A. M., increase of leprosy, 22; pamphlet on leprosy and vaccination, 205.
Brown, Mr. H., of Simla, on leprosy and vaccination, 189.
Browne, Dr., on leprosy and vaccination, 154.
Buckmaster, Dr. G. A., appointment of, as commissioner, 296.
Bulkeley, Dr. L. D., on contagion, 87.
Bullock, The Rev. G. M'Callum, statement by, 73.
Burma, leprosy in, 77.

Calcutta Daily News on vaccination and leprosy, 184.
Canada, leprosy in, 45, 325.
Cantharidin treatment, cruelty of, 346.
Cape Argus on increase of leprosy, 70.
Cape Colony, leprosy in, 65; medical reports from, 268, 275.
Carter, Dr. Vandyke, leprosy in Norway, 27; on contagion, 92; on vaccination and leprosy, 189.
Castor, Dr., opinion of, 50; on leprosy and vaccination, 174-5.
Census of lepers in India, 76, 256-61; inefficiency of, 261; in Leeward Islands, 262; Dr. Makuna's medical vaccination census, 376.
Ceylon, leprosy in, 78
Chambers's Encyclopædia, on leprosy and vaccination, 208.
Chaulmoogra oil, treatment by, 328.
Chew, Dr. Roger S., cases of inoculated and invaccinated leprosy, 191-9.

INDEX.

Clarke, Sir Andrew, 21.
Colombia, leprosy in, 46.
Commission, leprosy, to India, 296.
Committee, report of, Cape Colony. 68; Leprosy Investigation Committee appointed, 252; report of Leprosy Investigation Committee, 299.
Concealment of cases, 22-4, 41, 67, 72, 285-6, 289, 294
Constantinople, leprosy in, 37.
Contagion of leprosy, 64, 80, 96; personal inquiries into, 81; immunity of nurses from, 82; of attendants, 91, 93.
Contagion of leprosy, medical opinions on:—
Dr. A. Mouritz, 83.
Dr. Van Deventer, 84.
Dr. Trousseau, 84.
Dr. Manget, 84.
Dr. C. Handfield-Jones, 84
Dr Max Sandreczi, 84.
Mr. J. Hutchinson, 85.
Surgeon-Major Porteus, 86.
Dr. Van Someron, 86.
Dr. W. Monro, 86.
Report of Royal College of Physicians, 87
Sir Erasmus Wilson, 87
Dr. W. B. Atkinson, 87, 116.
Dr. Shoemaker, 87
Dr. L. D. Bulkley, 87.
Dr. H. M'Hatton, 87.
Dr J. L. Mears, 87.
Drs. Fox and Graham, 88.
Dr. J. W. Farrar, 88.
Dr. Beaven Rake, 88.
Mr. J. Freeland, 88.
Dr. J. H. Dunn, 89.
Dr. A. Ginders, 90.
Dr M. C. Ghose, 91.
Dr. Vandyke Carter, 92.
Dr. Day, 92.
Dr. J. Jackson, 92.
Mr. A. Mitra, 93.
Mr A. Mackenzie, 93.
Dr. W A Kynsey, 93.
Dr. Dixon, 94.
Dr. Ross, 95.
Dr. W. V. M. Koch, 95.

Contagion of leprosy, medical opinions on:—
Dr A Abercromby, 95.
Dr. Davidson, 95.
Cory, Dr. R., inoculation of, with syphilis, 213; on vaccinal diseases, 381.
Crabbe, Mr. H. G., on leprosy and vaccination, 166-7.
Creighton, Dr. C., researches by, 241, 387.
Crete, leprosy in, 37, 83.
Cultivation experiments, 344.
Cundimarea, leprosy in, 47.
Cunningham, Dr., 108.
Cures of leprosy considered, 310.
Currie, Dr. J. Z., on immunity of Indians, 267.

"DAILY GRAPHIC," letter in, 32; leprosy and vaccination, 145.
Dalrymple, Rev. W. D., a victim, 292.
Damien, Father, and contagion, 101, 121.
Danielssen, Dr., cases treated by, 323, 345.
Daubler, Dr., cases of leprosy and vaccination, reported by, 209-13
Davidson, Dr., on contagion, 95.
Day, Dr., on contagion, 92.
Dayton, Mr., information of, on leprosy and heredity, 230.
Deventer, Dr. Van, on contagion, 84.
Diagnosis, difficulty of, 232, 260.
Diet as a cause of leprosy, 225-6.
Distribution of leprosy, 24.
Dixon, Dr., on Contagion, 94.
Donovan, Dr., on cures, 321.
Drugs, uselessness of, 313.
Duckworth, Sir Dyce, on lepers, 304.
Dunn, Dr. J. H., on contagion, 89; on treatment, 326.
Dutch Guiana, leprosy in, 52.

EARLY indications of leprosy, 156.
Ebden, Dr. H. A., on leprosy, 64.
Egypt, leprosy in, 38.
Elephantiasis anæsthetica, case of, 203.

Elephantiasis tuberculosa, case of, 202.
Emerson, Dr., on increase of leprosy, 42.
England, leprosy in, 62, 85.
Escape, a narrow, 150.
Etiology of leprosy, Dr. Kaurin on, 157.
Experimental research, 313, *et seq.*

FARQUHAR, Dr. F., on leprosy and vaccination, 186.
Farrar, Dr. J. W., on contagion, 88.
Fear of vaccination, 251.
Fish as a cause of leprosy, opinion of Sir W. Moore, 225; of Mr. J. Hutchinson, 227; of Dr. A. Mitra, 228; of Dr. Hercules MacDonnell, 228; of Dr. C. N. Macnamara 229; of Dr. Murray MacLaren, 229; of Dr. J. Goldschmidt, 229.
Fitch, Dr, on heredity and leprosy, 224; on segregation, 283.
Forné, Dr, report on leprosy in New Caledonia, 57.
Fox, Dr. Colcott, cases of leprosy in United Kingdom, 62-3; on contagion, 88.
Fox, Dr. Tilbury, on leprosy and vaccination, 186.
France, leprosy in, 59.
Freeland, Mr. J., on contagion, 88; on leprosy and vaccination, 147.
Funeral ritual on lepers, 311.

GAIRDNER, Prof. W. T., on cases of leprosy induced by vaccination, 139-144.
Georgetown, hospital at, 50-52.
Gertrude, Sister Rose, 44.
Ghose, Dr. M. C., on contagion, 91; on vaccination and leprosy, 190.
Gibson, His Excellency Walter M., letter of inquiry, 247.
Ginders, Dr. A., on contagion, 90.
Goldschmidt, Dr. Julius, on leprosy in Madeira, 229, 373.
Graham, Dr., on contagion, 88; on immunity of Indians, 265.
Greek Archipelago, leprosy in, 37.

Greene, Dr., statement of, 38.
Grieve, Dr. Robert, on leprosy and vaccination, 175.
Gronvold, Dr. C., on heredity and leprosy, 224.
Gurjun oil, treatment by, 324-5, 330, 341.

HANSEN, Dr. G. A., on contagion, 80.
Hawaii, leprosy in, 42; increase of leprosy in, 246; attempted cures in, 314.
Hayes, Miss Alice, work of, 294.
Heidenstam, Dr., on leprosy and inoculation, 99.
Hellat, Dr. P., leprosy in Baltic Provinces, 26, 221; on vaccination and leprosy, 215.
Henry, Mr. Alexander, on leprosy and vaccination, 151-2.
Heredity and leprosy, Dr. Blanc on, 39; further considered, 223, *et seq.*
Hicks, Mr. E. H., on increase of leprosy, 48.
Hillebrand, Dr. W., on hereditary taint, 223.
Hillis, Dr. J. D., report of, 51; cases of leprosy due to vaccination, 169-171; letter of, 173-4.
Hoggan, Dr. George, letter on leprosy and vaccination, 201.
Honiball, Dr. O. D., on gurjun oil, 324.
Hospital Gazette, on leprosy and vaccination, 207.
Human beings inoculated with leprosy and syphilis, see leprosy and inoculation.
Hutchinson, Mr. J., on contagion, 85; on leprosy and diet, 227; notes on acquired leprosy, 235.
Hyde, Prof., on anæsthetic leprosy, 325.
Hygienic treatment the only palliative, 310.

ICELAND, leprosy in, 27.
Immunity of nurses, 82; attendants, 91-3; of aborigines, 264-7.

INDEX.

Impey, Dr. S. P., on leprosy and vaccination, 213 ; experience of, 281.
Inadequacy of medical theories of causation, 231-6.
Increase of leprosy, 21-79.
Incubation of leprosy, 155, 200.
Incurability of leprosy, 310.
India, leprosy in, 73, 182; leper census, 256-8.
Indians, freedom from leprosy, 264-7.
Indian Spectator on vaccination and leprosy, 185.
Indies, West, leprosy in, 28 ; Mr. Gourlay, M. P., and Sir W. Robinson, 34.
Influenza in Robben Island, 278.
Inoculation of leprosy, see leprosy and inoculation.
Insanitation, 154.
Introduction of leprosy by vaccination, see leprosy and vaccination.
Invaccinated diseases, 268-275.
Invaccination of leprosy, see leprosy and vaccination.
Isolation of lepers, 282.

JACKSON, Dr. J., 92.
Jamaica, leprosy in, 34, 321
Japan, leprosy in, 35 ; Dr. Albert S. Ashmead on, 129 ; Father Testevuide in, 290.
Jerusalem, leprosy in, 292.
Jones, Dr. Handfield, on contagion, 84.
Journal of Leprosy Investigation Committee, 230-1.

KAHOOKANO, Mr., bill to repeal vaccination laws in Hawaii, 166.
Kalihi, visit to, 43.
Kapiolani, Queen, visit of, to Molokai, 160.
Kashmir, leprosy in, 228.
Kaurin, Dr., on Leprosy, 157.
Keanu, inoculation of the convict, 123-8
Kimiball, Dr. J. H., opinion of, 41
Koch, Dr. W., report of, 30; on contagion, 95.
Kumaon, Leprosy in, 73.

Kynsey, Dr., leprosy in Ceylon, 78; on contagion, 93.

LACARY, Dr. N. (Guadaloupe), on difficulty of leper enumeration, 35
Lancet, on vaccination in Japan, 36 ; on leprosy in London, 85 : on contagion of leprosy, 86 ; Dr Wilson's communication in, 129 ; on heredity and leprosy, 224.
Lanoline, 182.
Larder, Dr., cases of leprosy, 63.
Leeward Islands, leper census in, 262.
Legrand, Dr. M A, leprosy in New Caledonia, 56; on barbarous therapeutics, 58.
Leicester and vaccination, 382.
Leloir, Dr. H., 237.
Leper census, see census.
Lepers, mendicant, 33 ; as tradesmen, 33 ; in society, 51 ; curing fish, 70 ; incarceration of, 71 : amongst white population, 76.
Leprosy, increase of, 21 ; medical opinion, 25; latency of anæsthetic, 179, 221 ; tubercular, 179, 221 ; in children, 49, 50, 55 : Dr. Sutliff on, 159 180; repression act in South Africa, 65, 71 ; *Cape Times* on, 287 ; contagion of, considered, 80, 95 ; Dr. White on contagion, 116; virus of, 81; Investigation Committee, 295, 309; and revaccination, 216 222 ; and diet, 225, 226; and the aboriginal races, 264-7 ; connection with mercury, 334.
Leprosy and inoculation, dread of, 83 ; chapter on, 97-130 ; inoculation of hospital attendants, 102; accidental inoculation, 103 ; Dr. W. B. Atkinson on, 116 : the convict Keanu, 123-8; Dr. C. N. Macnamara on, 125 ; inoculation by insects. :28-9 ; of lepers with syphilis, 348.
Leprosy caused by vaccination— Mr. Tebb's evidence, 7-10, 62.
In Japan, 36.
In Hawaii, 43, 160.
In South America, 48.

Leprosy and vaccination—
　In Venezuela, 53.
　In French Colonies, 58.
　In South Africa, 72.
　Connection between, 130.
　The question considered, 131-215.
　Sir Ranald Martin on, 134, 172.
　Royal Gazette, Trinidad, 134.
　Governor Robinson's circular, 136.
　Medical opinions on—
　　Dr. Alston, Dr. Chittenden, Dr. de Verteuil, Dr. R. F. Black, Dr. R. H. Knaggs, Dr. Beaven Rake, Dr. de Montbrun, 137.
　Letter of Dr. Black, 138.
　Professor Gairdner's cases, 139-144.
　Daily Graphic, 145.
　Dr. C. Burgoyne Pasley, 144-5.
　Public opinion on, in Trinidad, 146.
　Dr. John Freeland, 147.
　Dr. Taylor's letters, 147-9.
　Mr. Alexander Henry's evidence, 151.
　Surgeon J R. Tryon, 160.
　Native dread of vaccination, 153, 180.
　Drs. Bowerbank and Browne, 154.
　Dr. E. Arning on, 156-7.
　Mr. G. C. Potter, 161.
　H. G. Crabbe, 166-7.
　Dr. John D. Hillis's cases, 169; 171-4.
　Dr. Robert Grieve, 175.
　Dr. Castor, 174-5.
　Mr. Walter M. Gibson, Dr. H. S. Orme, 162, 177.
　Dr. H. G. Piffard, 171-2.
　Case of invaccinated leprosy, 178.
　Dr. Blanc, 179.
　Dr. R. B. Williams, 180.
　Dr. S. N. Boral, 182.
　The Statesman (Calcutta), 183.
　Calcutta Daily News and *Bombay Guardian*, 184.
　Indian Spectator, 185.
　Dr. Tilbury Fox, 186.
　Dr. Pringle's address on, 187.
　Dr. V. Carter and Mr. H. Brown, 189.

Leprosy and vaccination—
　Dr. Chunder Ghose, 190.
　Dr. Roger Chew's cases, 191-9.
　Dr. G. Hoggan's letter, 201.
　Dr. Suzor, 204.
　Mr. Malcolm Morris, 205.
　Dr. A. M. Brown, 205.
　Dr. Bechtinger and Dr. W. Munro, 207.
　Hospital Gazette, 207.
　Occidental Medical Times, 208.
　Chambers's Encyclopædia, 208.
　Dr. George Thin, 208.
　Dr. Daubler's cases, 209-213.
　Dr. Impey, 213.
　Dr. P. Hellat, 215.
　Danger to recruits, 221.
　International Hygienic Congress, 236-55.
　Governor Longden's dispatch, 242
　Dr Reade, 244.
　A Government Commissioner on, 264.
　Connection between leprosy and vaccination in South Africa, 268-275.
　Interrogation of lepers at Robben Island, 280.
　The author's summary, 389.
Lima, Dr., on leprosy in Brazil, 231-3; on immunity of Indians, 266.
Livonia, leprosy in, 218.
Longden, Governor, on leprosy and vaccination, 242.
Lovell, Dr. T., circular to practitioners, 248.
Lymph, animal, etc., 181-2, 213.

M'Cormick, leprosy in Iceland, 27
M'Hatton, Dr. H., on contagion, 87.
Mackenzie, Mr. A., on contagion, 93.
Mackenzie, Sir Morell, 22; and Dr Boon, 32.
MacLaren, Dr. Murray, on leprosy in New Brunswick, 229.
Macnamara, Dr. C N., on leprosy and inoculation, 109, 125; on leprosy and diet, 228.

INDEX. 399

M'Turk, Mr., on leprosy and vaccination, 264.
Madagascar, leprosy in, 45.
Madeira, leprosy in, 229 ; Dr. Goldschmidt on, 373.
Mahaica Creek, 51.
Mahommedan repugnance to vaccination, 100.
Main, leprosy in, 23.
Malling, Mr. J. C., report of, 34.
Manakeke, isle of, leprosy in, 55.
Manget, Dr., on contagion, 84.
Maoris, leprosy among, 120.
Maracaibo, Consul Plumacher's report from, 52 ; visit to, 292.
Maritime Medical News, 229.
Marsden, Miss Kate, 293.
Martin, Sir Ranald, on danger of vaccination, 172.
Mauritius, leprosy in, 45.
Mears, Dr. J. L., on contagion, 87.
Medical evidence as regards communicability of leprosy by inoculation : -
Sir W. Moore (Bombay), 97.
Dr. Heidenstam (Cyprus), 98.
Dr Olavide (Madrid), 99.
Dr. Zambaco-Pacha(Mitylene),99.
Dr. Sutherland (Patna), 100.
Dr. Forné (West Indies), 101.
Dr. Woods, 102
Professor Cayley, 104.
Dr. Thin, 104.
Dr. John Murray, 104.
Dr. Liveing, 104.
Dr. G A. Hansen, 105.
Sir E. Wilson, 105.
Dr. B. Krishna, 106.
Surgeon-Major Pringle, 107.
Dr Joq. F. Periera, 107.
Dr. Cunningham, 108.
Dr. Shircore, 108.
Dr. Ackworth, 108.
Dr. Hatch, 108.
Dr. Neve, 108.
Dr. G. D. M'Reddie, 108.
Surgeon C. N. Macnamara, 109
Hon. Dr. Atherstone, 110.
Dr. Abercrombie, 110.
Sir Samuel Needham, 111.
Dr. W. H. Ross, 112.

Medical evidence as regards communicability of leprosy by inoculation :—
Dr. J. C. Taché, 113.
Dr. Manget, 114.
Dr. C. W. Allen, 114.
Dr. H. S. Orme, 115.
Dr. S. Kneeland, 115.
Dr. T. C. Graham, 115.
Dr. W. B. Atkinson, 116.
Dr. White, 116.
Dr. R. J. Farquharson, 117.
Dr. Wood (U. S. Navy), 118.
Dr. T. H Bemiss, 118.
Dr. Hall Bakewell, 119, 132-5.
Dr. N B. Emerson, 119.
Dr. J. D. Hillis, 119.
Dr. A Mowich, 120.
Dr. T. J. Kimball, 120.
Dr. Ginder, 120.
Dr. W. Munro, 121.
Dr. J. Freeland, 122.
Medicinal agents, 313, *et seq*.
Meyer, Mr., report of, 42 ; mercury and leprosy, 334.
Milroy, Dr. Gavin, on leprosy and vaccination, 153 ; inquiries of, 243.
Mitra, Dr. A., on contagion, 92 ; on leprosy and diet, 228.
Mitylene, leprosy in, 37, 100.
Mixed tuberculated leprosy, 221.
Molokai, leprosy in, 24 ; deportation of lepers to, 44 ; Father Damien in, 101 ; Mr. Crabbe, 167 ; Dr. George L. Fitch, 283.
Moore, Sir W., on leprosy and fish diet, 225 ; on leprosy, syphilis, and inoculation, 98.
Moravian Brothers, devotion of, 291.
Morrow, Dr. Prince A., opinion of, 39.
Mosquitoes and leprosy, 129.
Mouritz, Dr. A , on contagion, 83.
Münch, G. N., leprosy in S. Russia, 25.
Munro, Dr. (St. Kitts), 31 ; on contagion, 86 ; on leprosy and vaccination, 207.

NATIONAL Leprosy Investigation Committee, 295.

Native prejudice against vaccination, 100, 155, 180.
New Brunswick, leprosy in, 229.
New Caledonia, leprosy in, 56.
New Orleans, leprosy in, 39.
New South Wales, leprosy in, 234.
New Zealand, leprosy in, 54.
Nineteenth Century, Sir. M. Mackenzie's paper in, 22.
Norway, leprosy in, 27.

Occidental Medical Times on leprosy and vaccination, 208.
Oceana, leprosy in, 55.
Official inquiries into leprosy, 241; bias concerning vaccination, 254·5; courtesy, 164; statistics, 256-8.
Olavide, Dr. (Madrid), on leprosy and inoculation, 99.
Orme, Dr., on secretion of lepers, 24; on leprosy and vaccination, 162, 177; on treatment of leprosy, 326.

PAGET, Sir James, on surgical pathology, 375.
Palliative treatment, Dr. Hillis on, 324.
Pasley, Dr. C. Burgoyne, on leprosy and vaccination, 144-5.
Perron, Dr., on tubercular vaccination, 357.
Physicians, Royal College of, report of, to Lord Kimberley, 155.
Piffard, Dr. H. G., on leprosy and vaccination, 171·2.
Plumacher, Consul F. H., reports of, 52; on inoculation, 117; on treatment of leprosy, 328.
Porteus, Surgeon-Major, on contagion, 86.
Pringle, Surgeon-Major, on leprosy and vaccination, 187-8.
Public Health Act (Cape of Good Hope), 268.

RACE, influence of, 223.
Rake, Dr. Beaven, increase of leprosy in Trinidad, 30; on contagion, 88.

Reade, Dr, on leprosy and vaccination, 244.
Reports, Royal College Physicians', 87, 249; report in reply to Mr. Gibson's circular, 247; report on leprosy and yaws, 245; of Leprosy Investigation Committee, 299.
Revaccination and leprosy, 216, 222.
Revolt against compulsory vaccination, 217; in India, 360.
Riga, leprosy in, 26.
Robben Island, visit to, 276, 281; report from, 288.
Robinson, Governor Sir W, on increase of leprosy, 34; circular of, 136.
Ross, Dr. W. H., evidence of, 68; on contagion, 95.
Royal commission on vaccination, see dedication.
Russia, leprosy in, 24.

SAGER, Mr. H., on causation of leprosy, 234.
St. Kitts, leprosy in, 31; *Lazaretto* on the leper census, 262.
St. Vincent, leprosy in, 33.
Samos, leprosy in, 36.
Sandreczi, Dr. Max, 84; on leprosy in Palestine, 230.
Sandwich Islands, leprosy in, 39, 156, 157.
Sanitation, the great prophylaxis, 325.
Santander, leprosy in, 46.
Segregation, difficulty of, 40; consideration and failure of, 282-289; report of Leprosy Investigation Committee, 301.
Self-devotion to lepers, 290-4.
Shoemaker, Dr., on contagion, 87.
Siler, Consul, report of, 69; on inoculation, 117.
Simon, Sir John, 365.
Small-pox, outbreaks of, 43.
Smeaton, Mr., leprosy in Burma, 77.
Smith, Dr. A. C., on treatment of leprosy, 327.
Soldiers and leprosy, 217-220.
Someron, Dr. Van, on contagion, 86.

Spain, leprosy in, 59.
Sprigg, Sir Gordon, enumeration of lepers by, 65.
Statesman, The Calcutta, 183.
Stubbs, Dr. T, report of Dr. Daubler's cases, 211-213.
Summary, 350.
Sutherland, Dr. (Patra), on leprous taint, 218.
Sutliff, Dr. F. B., leprosy in Hawaii, 23; in children, 159; case of invaccinated leprosy, 178.
Suzor, Dr., of Mauritius, on leprosy and vaccination, 204.
Syphilis and vaccination, 149-50.
Syria, leprosy in, 37.

TACHÉ, Dr. J. C., on treatment, 327.
Taylor, Dr. C. T., letters on leprosy and vaccination, 147-9.
Tebb, Mr. W., letter sent to *Lancet* by, 238; letter to Dr. Abraham, 253; personal statement, 389-391; visit to Robben Island, 276; summary concerning vaccination, 389, *et seq.*
Testevuide, Father, devotion of, 291.
Therapeutics, uselessness of, 325.
Thin, Dr. George, work on leprosy, 52; table of cases in Spain, 60, 61; on leprosy and vaccination, 208.
Tocaima, leprosy in, 89.
Tolima, leprosy in, 47.
Tonquin, leprosy in, 79.
Tracadie, leprosy in, 229.
Treatment of leprosy, 310-328, *et seq.*
Trinidad, leprosy in, 30.
Trousseau, Dr., on contagion, 84.
Tryon, Surgeon J. R, U.S. Navy, on leprosy and vaccination, 160.
Tubercular leprosy, see leprosy.
Tuberculin, treatment of leprosy by, 339.
Tuberculosis, danger concerning, 355, *et seq.*

Turner, Dr W. R., on leprosy in Saldanha Bay, 64.

VACCINAL diseases, 268, 275; Dr. Bakewell on, 371
Vaccination, a cause of leprosy (see leprosy and vaccination); as a source of danger ignored, 254; frauds in India, 354; Mr Tebb's summary concerning, 389, *et seq.*
Valladores, Dr. Miguel, on immunity of Indians, 265; on mercurial treatment, 329.
Value of hygienic measures, 320.
Veendam, report of Dr. J. L, 50.
Venezuela, leprosy in, 52.
Verteuil, Dr. de, on leprosy and vaccination, 137
Village, a leper, near Hanoi, 79.
Virus, leper, 81; humanised, to be avoided, 179; danger of animal, 381.

WALES, Prince of, 21; speech at Marlborough House, 73.
West Indies, see Indies, West.
Westminster Review, papers by Mr. Tebb in, 241.
Wheeler, Mr. T H., report of, 46; on contagion, 89.
White, Dr., report of, 41; on contagion of leprosy, 116.
Williams, Dr. R B, on leprosy and vaccination, 180.
Wilson, Sir Erasmus, on contagion, 87
Wright, Dr. H. J., evidence of, 331.

YAMAMOTO, Dr. K., 338.
Yaws, report on leprosy and, 245.

ZAMBACO, Dr., leprosy in Eastern Europe, 37; on leprosy in Mitylene, 99.
Zeidan, Dr. Selim, on leprosy in Egypt, 38.
Zouaves, regiment of, syphilised by vaccination, 185.

LIST OF WORKS AND PAMPHLETS REFERRED TO IN THE FOREGOING PAGES.

A CLINICAL LECTURE ON ANÆSTHETIC LEPROSY, by James Nevins Hyde. Chicago, December 12th, 1878.

AMERICAN JOURNAL OF THE MEDICAL SCIENCES.—Edited by Edward P. Davis. Philadelphia, July, 1891. Vol. cii., No. 1.

AN ABSTRACT OF LECTURES ON LEPRA, by J. L. Bidenkap.—Christiania, 1886.

APPENDIX TO THE REPORT ON LEPROSY OF THE PRESIDENT OF THE BOARD OF HEALTH TO THE LEGISLATIVE ASSEMBLY OF 1886.—Honolulu.

ARCHIVES DE MÉDECINE NAVALE ET COLONIALE.—Éditeur, Octave Doin. Paris, Septembre, 1890.

ARCHIVES DE MÉDECINE NAVALE ET COLONIALE.—Éditeur, Octave Doin. Paris, Février, 1891.

A REMARKABLE EXPERIENCE CONCERNING LEPROSY, by W. T. Gairdner, M.D., reprinted from the *British Medical Journal*, June 11, 1887.

BIENNIAL REPORT OF THE BOARD OF HEALTH TO THE GENERAL ASSEMBLY OF THE STATE OF LOUISIANA, 1888 AND 1889.—C. P. Wilkinson, M.D., President. 1890.

BIENNIAL REPORT OF THE PRESIDENT OF THE BOARD OF HEALTH OF THE LEGISLATURE OF THE HAWAIIAN KINGDOM.—Session of 1888. Honolulu, 1888.

BIENNIAL REPORT OF THE PRESIDENT OF THE BOARD OF HEALTH OF THE LEGISLATURE OF THE HAWAIIAN KINGDOM.—Session of 1890. Honolulu, 1890.

BOSTON MEDICAL AND SURGICAL JOURNAL.—May 15, 1884. Boston, Mass.

CAPE COLONY (SOUTH AFRICA) DISTRICT SURGEONS' REPORTS—1884-1891.

CHICAGO MEDICAL JOURNAL AND EXAMINER. April, 1882.

COLONY OF MAURITIUS.—Report of the Leprosy Inquiry Commission and Annexures. 26th October, 1888.

COMPULSORY VACCINATION.—Copy of a Letter, dated 30th June, 1855, addressed to the President of the Board of Health, by John Gibbs, Esq., entitled, "Compulsory Vaccination, briefly considered in its Scientific, Religious, and Political Aspects." (Mr Brotherton.) Printed 31st March, 1856.

COPIES OF REPORT OF DR. EDWARD ARNING TO THE BOARD OF HEALTH, AND CORRESPONDENCE ARISING THEREFROM.—1885.

CORRESPONDENCE RELATING TO THE DISCOVERY OF AN ALLEGED CURE OF LEPROSY.—Presented to both Houses of Parliament by command of Her Majesty. May, 1871.

COW-POX AND VACCINAL SYPHILIS, by Dr. Charles Creighton.

CYPRUS.—Report by Dr. Heidenstam, C.M.G., Chief Medical Officer on Leprosy in Cyprus. March, 1890.

DEDICATION OF THE KAPIOLANI HOME.—Honolulu, Nov. 9, 1885.

DIE LEPRA AUF MADEIRA, von Dr. Julius Goldschmidt.—Leipzig, 1891.

ETIOLOGY OF LEPROSY, by George L. Fitch, M.D. New York.

JENNER AND VACCINATION, by Charles Creighton, M.D. 1889.

JOURNAL OF THE LEPROSY INVESTIGATION COMMITTEE, edited by Phineas S. Abraham —No. 1, August, 1890. London.

JOURNAL OF THE LEPROSY INVESTIGATION COMMITTEE, edited by Phineas S Abraham —No. 2, February, 1891. London.

JOURNAL OF THE LEPROSY INVESTIGATION COMMITTEE, edited by Phineas S. Abraham.—No. 3, July, 1891. London.

JOURNAL OF THE LEPROSY INVESTIGATION COMMITTEE, edited by Phineas S. Abraham.—No. 4, December, 1891. London.

LEPROSY, by W. Munro, M.D., C.M. Manchester, 1879.

REFERRED TO IN THE FOREGOING PAGES. 405

LEPROSY, by George Thin, M.D. London, 1891.

LEPROSY: A COMMUNICABLE DISEASE.—C. N. Macnamara. London, 1889.

LEPROSY AND LEPER HOUSES, by Sir William Moore. (The Hospitals Association Pamphlets. No. 38.)

LEPROSY: AN IMPERIAL DANGER.—H. P. Wright, M.A. London, 1889.

LEPROSY: A Review of some facts and figures.—A paper read before the Epidemiological Society of London, May 8th, 1889, by Phineas S. Abraham. London, 1889.

LEPROSY CONTAGIOUS: Extracts from Report of Royal College of Physicians, 1867.

LEPROSY, CONTAGIOUSNESS, ETC.—Extracts from Reports and Evidence Select Committee, Cape Town House of Assembly, 1883 and 1889.

LEPROSY: Correspondence, etc., relating thereto. Trinidad, 1890. (Government Printing Office.)

LEPROSY IN BOMBAY, IN ITS MEDICAL AND STATE ASPECTS, by Balchandra Krishna.

LEPROSY IN BRITISH GUIANA.—An Account of West Indian Leprosy, by John D. Hillis. London, 1881.

LEPROSY IN FOREIGN COUNTRIES.—Summary of Reports. Honolulu, 1886.

LEPROSY IN HAWAII.—Extracts from Reports of Presidents of the Board of Health, Government Physicians, and others, and from Official Records, in regard to Leprosy before and after the passage of the Act to prevent the spread of Leprosy, approved January 3, 1865. Supplement. The laws and regulations in regard to Leprosy in the Hawaiian kingdom. Honolulu, 1886.

LEPROSY IN MINNESOTA.—Report to the State Board of Health of Minnesota, by Dr. Chr. Gronvold, reprinted from *Chicago Medical Journal and Examiner* for February, 1884.

LEPROSY; ITS EXTENT AND CONTROL, ORIGIN AND GEOGRAPHICAL DISTRIBUTION, by H. S. Orme, President, State Board of Health, California.

27

MADRAS ACT, No. 1, of 1884; THE CITY OF MADRAS MUNICIPAL ACT, 1883.

MARINE HOSPITAL SERVICE OF THE UNITED STATES.—Annual Report of the Supervising Surgeon-General, Washington, U.S., 1891

MARITIME MEDICAL NEWS.—Vol. ii., No. 3, May, 1890. Halifax, N.S.

MARITIME MEDICAL NEWS.—Vol. ii., No. 4, July, 1890. Halifax, N.S.

MEDICAL NEWS.—Philadelphia, Dec. 17, 1887. Vol. li. No. 25.

MONITOR OF ELECTRO-HOMŒOPATHY, N. I. (third year). — 30th August, 1891. Bologna.

MY LEPER FRIENDS, by Mrs. M H. Hayes, with chapter on Leprosy by Surgeon-Major G. G. Maclaren, M.D. London. 1891.

NEW YORK MEDICAL JOURNAL.—Vol. xlvii, No. 12. March 24, 1888.

NEW YORK MEDICAL JOURNAL.—Vol. xlvii., No. 13. March 31, 1888.

NORTH AMERICAN REVIEW. June, 1891.

NORTH-WESTERN LANCET.—Vol. viii., No. 5. St. Paul, Minn., March 1, 1888.

OCCIDENTAL MEDICAL TIMES.—Vol. iii., No. 4, April, 1889. Sacramento, California.

OCCIDENTAL MEDICAL TIMES.—Vol. iv., No. 4, April, 1890. Sacramento, California.

OCCIDENTAL MEDICAL TIMES.—Vol. iv., No. 6, June, 1890. Sacramento, California.

OCCIDENTAL MEDICAL TIMES.—Vol. iv., No. 9, September, 1890. Sacramento, California.

OCCIDENTAL MEDICAL TIMES.—Vol. iv., No. 10, October, 1890. Sacramento, California.

OCCIDENTAL MEDICAL TIMES.—Vol. v., No. 1., January, 1891. Sacramento, California.

ON LEPROSY AND ELEPHANTIASIS, by H. Vandyke Carter, M.D., H. M. Ind. Med. Service. London, 1874.

PERSONAL OBSERVATIONS OF LEPROSY IN MEXICO AND THE SANDWICH ISLANDS, by Prince A Morrow, M.D. Reprinted from the *New York Medical Journal* for July 27th, 1889.

REPORT FOR THE YEAR 1889 ON THE TRADE OF COLUMBIA. Foreign Office. 1890.

REPORT OF MEDICAL OFFICER OF THE LEPER ASYLUM, MAHAICA, BRITISH GUIANA, FOR 1880.

REPORT OF THE PRESIDENT OF THE BOARD OF HEALTH TO THE LEGISLATIVE ASSEMBLY OF 1886.—Honolulu, 1886

REPORT ON LEPROSY AND YAWS IN THE WEST INDIES, addressed to Her Majesty's Secretary of State for the Colonies, by Gavin Milroy, M.D. 1873.

REPORT ON THE ETIOLOGY OF LEPROSY TO THE CALIFORNIA STATE MEDICAL SOCIETY, by W. F. M'Nutt, M.D.

REPORTS FROM THE CONSULS OF THE UNITED STATES ON THE COMMERCE, MANUFACTURES, ETC., OF THEIR CONSULAR DISTRICTS - No. 10 August, 1881.

REPORTS FROM THE CONSULS OF THE UNITED STATES.—No. 79. June, 1887. Washington, U.S.A.

REPORTS FROM THE CONSULS OF THE UNITED STATES—No. 115. April, 1890. Washington, U.S.A.

REPORTS FROM THE CONSULS OF THE UNITED STATES—No. 119. August, 1890. Washington, U.S.A.

REPORTS OF THE MEDICAL COMMITTEE, THE VACCINATING SURGEON, THE INSPECTOR OF ASYLUMS, ETC., CAPE OF GOOD HOPE, FOR 1890

SCHEME FOR OBTAINING A BETTER KNOWLEDGE OF THE ENDEMIC SKIN DISEASES OF INDIA, by Tilbury Fox, M.D., and A. Farquhar, M.D. 1872.

SOME COMMENTS ON LEPROSY IN ITS CONTAGIO-SYPHILITIC AND VACCINAL ASPECTS, by A M. Brown, M.D. 1888.

SUPPLEMENT TO THE "JAMAICA GAZETTE" (New Series). - Vol. xiv., No. 10. June 18th, 1891.

THE ETIOLOGY OF LEPROSY: a Criticism of some current views, by P. S. Abrahams. Reprinted from the *Practitioner*.

THE GERMICIDE, edited by R. R. Russell, M.D.—Vol. vi. Nov., 1891. New York.

THE INDO-EUROPEAN CORRESPONDENCE, Calcutta.—Vol. xxvi, No. 39. September 30th, 1891.

THE LAZARETTO.—Basseterre, St. Kitts, B.W.I., 1890 to 1892.

THE NATIONAL LEPROSY FUND. London, 1890.

THE PRACTITIONER.—Edited by T. Lauder Brunton, Donald Macalister, and T. Mitchell Bruce. London, April, 1890. Vol. xliv., No. 4.

THE SANITARIAN.—Vol. xiii., No. 176, July, 1884. New York.

THE SPREAD OF LEPROSY, by John D. Hillis, F.R.C.S.—Reprinted from *Timehri*. Demerara, 1889.

TRAITÉ PRATIQUE ET THEORETIQUE DE LA LÈPRE.—Par Henri Leloir. 1886.

TRINIDAD ROYAL GAZETTE.—August 15, 1888. Vol. lvii.

TUBERCULAR LEPROSY.—By Alexander Abercromby, M.D. Edinburgh, 1861.

VACCINATION.—Mortality Return relating to Births and Deaths in England and Wales, Vaccination, Small-pox, etc. (Mr. Hopwood). 14th August, 1877.

VALUE OF VACCINATION: being a Précis or Digest of Evidence taken, *viva voce* (1871) before a Committee of the House of Commons (on the Vaccination Act,) 1867, by T. Baker.

Hay Nisbet & Co., Printers, 25 Jamaica Street, Glasgow, and 169, Fleet Street, London.

www.ingramcontent.com/pod-product-compliance
Lightning Source LLC
Chambersburg PA
CBHW022116290426
44112CB00008B/696